2017 年"一流应用技术大学"建设系列教材

计算机辅助制造

——基于 UG NX 12.0 的典型零件数控加工

Computer-aided Manufacturing(CAM)—UG NX 12.0-based CNC
Manufacturing of Typical Parts

主 编 沈骏腾

副主编 李彦杰 刘华 吕万江 龚勋

U0169840

西安电子科技大学出版社

内 容 简 介

本书结合作者多年从事 UG CAM 的教学和企业生产加工经验，以目前最新版本 UG NX 12.0 为操作平台，详细介绍了 UG CAM 三轴加工中心的平面铣、曲面铣、孔加工、仿真加工、后置处理等功能在实际生产中的应用。书中采用文字和图形相结合的形式，详细介绍了 5 个典型加工案例的数控加工工艺和 NX 12.0 软件的操作步骤。本书配套操作过程的动画演示、重要知识点的微课等教学资源，可以帮助读者更加直观地掌握 UG NX 12.0 软件的界面和操作步骤。

本书共分为 5 个项目，分别为开关盒中框的加工程序编制、批量加工零件的工艺设计与程序编制、自行车尾灯注塑模具动模型芯的程序编制、自行车尾灯注塑模具定模型芯的程序编制及吹塑模具瓶体的程序编制。

本书可作为 CAM 专业课程教材，特别适合 UG 软件的初、中级学者及各高等院校机械、模具、机电及相关专业的师生教学、培训和自学使用，也可作为企业从事数控加工、自动编程的广大工程技术人员的参考书。

图书在版编目(CIP)数据

计算机辅助制造：基于 UG NX 12.0 的典型零件数控加工 / 沈骏腾主编. —西安：
西安电子科技大学出版社，2020.6(2023.1 重印)
ISBN 978–7–5606–5189–7

Ⅰ. ① 计⋯   Ⅱ. ① 沈⋯   Ⅲ. ① 计算机辅助制造   Ⅳ. ① TP391.73

中国版本图书馆 CIP 数据核字(2018)第 278307 号

策　　划　毛红兵　万晶晶
责任编辑　万晶晶
出版发行　西安电子科技大学出版社(西安市太白南路 2 号)
电　　话　(029)88202421　88201467　　　邮　　编　710071
网　　址　www.xduph.com　　　　　　　电子邮箱　xdupfxb001@163.com
经　　销　新华书店
印刷单位　陕西天意印务有限责任公司
版　　次　2020 年 6 月第 1 版　　2023 年 1 月第 2 次印刷
开　　本　787 毫米×1092 毫米　1/16　印　张　21
字　　数　496 千字
印　　数　1001～2000 册
定　　价　69.00 元(含光盘)
ISBN 978–7–5606–5189–7 / TP
XDUP 5491001–2
如有印装问题可调换

天津中德应用技术大学

2017年"一流应用技术大学"建设系列教材

编 委 会

主 任：徐琤颖

委 员：（按姓氏笔画排序）

王庆桦　王守志　王金凤　邓　蓓　李　文

李晓锋　杨中力　张春明　陈　宽　赵相宾

姚　吉　徐红岩　靳鹤琳　薛　静

前　　言

　　计算机辅助制造(Computer-aided Manufacturing，CAM)是指在机械制造业中，利用电子数字计算机控制各种数字控制机床和设备，自动完成产品的加工、装配、检测和包装等制造过程。CAM 软件通常是指用于编制加工零件数控加工程序的软件平台。CAM 软件可以把图形化的三维模型转化为机床能够使用的 G 代码文件，加工出所需的各种零件。CAM 编程主要应用于三轴加工中心机床、多轴加工中心机床、数控车床、线切割机床等常见数控设备。现在市面上常用的 CAM 软件有 NX、Mastercam、Cimatron、Powermill、CAXA 制造工程师等。

　　本书由从事模具制造和数控加工工作的一线从业者编写。编者都具有十多年的企业生产经验，长期从事模具设计与制造、加工中心零件加工生产等工作，并且具有多年高职院校加工中心实践课和 CAM 软件教学的执教经验。编者通过多次的企业调研，发现大部分机加企业使用的 CAM 软件都是 UG，而且 UG 软件的使用范围比较广泛，包含企业中从产品设计、分析到加工等全方面的工作。经过多方面对比并考虑到学生毕业后的就业前景，我们在本书中选用 UG NX 12.0 作为教学软件，以便读者学习后能更快地融入企业。

　　本书以 UG NX 12.0 软件为平台，通过天津中西机床技术培训中心对外加工的五个真实案例，讲解了使用 NX 12.0 软件进行数控加工的方法和技巧。本书采用项目教学的形式，完整地展现了数控加工零件的零件结构分析、加工工艺设计、装夹胎具使用、刀具选择、加工程序编制、加工路径仿真、生成后处理代码等完整的 CAM 过程，讲述了 NX 12.0 软件在数控加工中的应用。全书共 5 个项目，包括开关盒中框的加工程序编制、批量加工零件的工艺设计与程序编制、自行车尾灯注塑模具动模型芯的程序编制、自行车尾灯注塑模具定模型芯的程序编制、吹塑模具瓶体的程序编制等加工案例。

　　本书每个项目都结合实例进行讲解。编者希望通过这样的内容安排，能够展现出 NX 12.0 数控加工的精髓，使读者看到完整的数控加工过程，进一步加深对 NX 12.0 数控加工模块的理解和认识，体会 NX 12.0 优秀的设计思想和设计功能，从而在以后的工作中熟练使用。

　　作为国际化专业建设用书，为满足外国友人来华学习专业课程知识的需求，本书在附录中用英文给出了本书项目一和项目五的重点内容。其中项目一介绍了平面轮廓类零件的加工程序编制，项目五介绍了曲面轮廓类零件的程序编制。通过两个重点章节英文内容的学习，可使外国读者基本掌握 NX 12.0 的数控编程方法。

　　本书含有学习光盘、微课、PPT 等多媒体教学资源。其中光盘中包含全书案例源文件、案例的编程后文件及后处理文件、练习等。书中重要知识点可通过扫描相应位置的二维码在线观看相应的微课讲解，每个项目最后的二维码包含本项目案例的操作视频讲解。

　　本书正文和光盘内容由沈骏腾、李彦杰、刘华编写完成；微课、动画内容由吕万江和李彦杰设计制作；英文章节的翻译和校对工作由龚勋完成。

　　本书在编写过程中得到了天津中德应用技术大学相关领导、天津中西机床技术培训中心、天津中西数控技术有限公司的大力支持，在此向他们表示衷心的感谢。

　　由于编者水平有限，书中难免有纰漏和不足之处，恳请广大读者、同仁不吝赐教。

编　者
2020 年 1 月

目　　录

项目一　开关盒中框的加工程序编制

案例说明 ✍

本项目以开关盒中框加工编程为案例，讲解开关盒中框的加工工艺、加工方法的选择、同一加工工序使用不同加工方法的对比、切削刀具的选择以及平面加工编程的注意事项等。

学习目标 ✍

通过学习开关盒中框零件的加工编程，读者应了解和掌握 UG NX 12.0 软件三轴平面零件的加工编程方法，并且能够举一反三，触类旁通。

学习任务 ✍

平面轮廓类零件是机械加工中常见的零件，其品种比较多，且大多是批量加工，多应用于电子、医疗、机械、航空等工业领域中。

本项目以开关盒中框加工编程为案例讲解 UG NX 12.0 软件三轴平面编程，重点是应学会并理解软件各种平面轮廓程序的加工方法和应用，并可以按照不同的方法和工艺生成加工程序。

图 1-1 是 UG NX 12.0 软件加工过程的流程图，它表明了创建和处理程序的过程和步骤。本书将以这个流程图作为引导进行讲解。

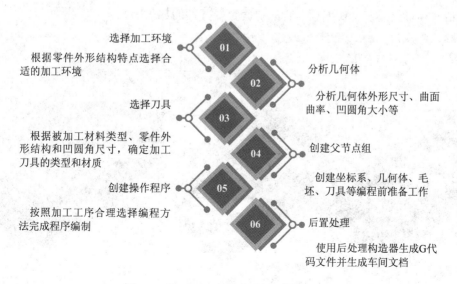

图 1-1　UG NX 12.0 软件加工过程流程图

1.1 开关盒中框的加工工艺规程

加工工艺规程描述了每步加工过程，一般包括加工区域、加工类型(平面铣、曲面铣、孔加工等)、工序内容、零件装夹方式、所需刀具及完成加工所必需的其他信息。

开关盒中框零件加工程序编制

开关盒中框的外形尺寸为 80 mm × 40 mm × 40 mm，材料为铝合金 6061。为便于批量加工并且降低加工成本，购买原材料时，一般选用外形尺寸稍大一点的毛料自行加工外形尺寸，不需要定制尺寸正好的六面精加工毛坯材料。因此，此次选用双边余量为 5 mm，尺寸为 85 mm × 45 mm × 45 mm 的矩形铝合金型材。下面的工艺过程简单描述了开关盒中框的加工过程，描述的内容包括工序号、顺序号、加工机床、工序内容、工序名和刀具名称。其定义如下：

工序号——工艺分布的顺序号。

顺序号——其中一个工艺下的加工顺序号。

加工机床——工序使用机床的选择。

工序内容——工序的详细内容描述。

工序名——工序加工特定的特征或任务的实际名称。

刀具名称——加工使用刀具的类型和材质。

1.1.1 案例工艺分析

图 1-2 所示为开关盒中框模型，此工件需要上下两方向加工，且装夹两次才能完成工件的制作。为保证加工后零件的尺寸精度和形位公差，在编程前首先要确定好工件的加工工序方案。

选择装夹加工方式

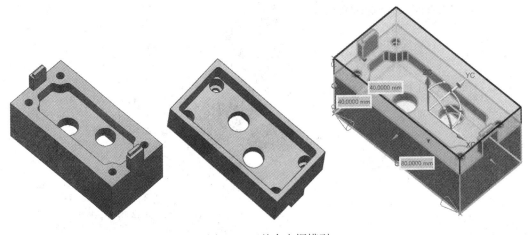

图 1-2 开关盒中框模型

方案一：

　　先使用虎钳装夹加工零件正面深度范围为 0～14.5 mm 的部分，然后夹住铣出来的两个凸起部分，再装夹下半部分，如图 1-3 所示。加工底部时由于装夹面较小，且加工材料为铝合金，材质较软，装夹力度过小加工时工件容易飞出，如果力度过大，那么工件很容易装夹变形，而且加工时肯定会产生很大的颤动，导致工件尺寸超差，且加工难度大、废品率高，不宜使用。

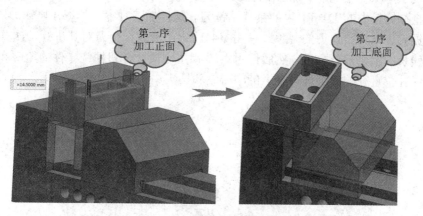

图 1-3　加工方案一示意图

方案二：

　　先使用虎钳装夹加工零件底面虎钳装夹深度为 10 mm 的部分，随后加工底部外形、型腔、孔等除正面两个凸台以外能加工的所有部分，然后夹住铣出来的矩形部分，再加工零件正面两凸台和型腔部分，如图 1-4 所示。使用这种加工方案在加工第二序时工件的装夹面积较大，而且需要加工的毛坯余量较小，且对零件的装夹力度没有太大的要求，从而能提高零件的装夹速度和加工零件的合格率，因此在本次加工中选用方案二。

项目一第一序装夹方式　　　　　　　　项目一第二序装夹方式

图 1-4　加工方案二示意图

1.1.2 案例加工刀具的选择

本例开关盒中框零件的加工材料为铝合金 6061，外形尺寸为 80 mm×40 mm×40 mm，最小凹圆角如图 1-5 所示，圆角半径为 2 mm。型腔内外轮廓均由平面构成。首先对其进行粗加工，粗加工时要去除大量的材料，加工时刀具承受的力会很大，所以要选取直径比较大的 D16R0.4 的镶片钻铣刀。这样加工时不容易发生振刀、断刀、崩刃的现象。使用 φ4 的铝用硬质合金刀对其进行二次开粗，清理粗加工后残余的圆角部分余量，然后使用 φ10 铝用硬质合金刀精铣零件的轮廓和底面，最后按照零件孔位直径的大小选择合适的中心钻和钻头加工零件上所有的孔。

常用切削刀具的介绍

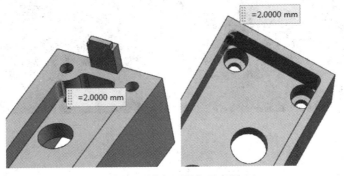

图 1-5　最小凹圆角示意图

按照上述零件加工工艺方案和切削刀具的选择方式，合理安排零件的加工工艺过程。按照先粗后精、先面后孔、基准统一的原则设计本案例的加工工艺过程单，如表 1-1 所示。

表 1-1　开关盒中框工艺过程单

工序号	顺序号	加工机床	工序内容	工序名	刀具名称
底面	1	加工中心	粗加工	型腔铣	D16R0.4
底面	2	加工中心	二次开粗	剩余铣	D6
底面	3	加工中心	底面精加工	底壁铣	D10
底面	4	加工中心	精加工侧壁	底壁铣	D6
底面	5	加工中心	精加工侧壁	精铣壁	D10
底面	6	加工中心	打中心钻	定心钻	中心钻 D6
底面	7	加工中心	加工 φ4.5 的孔	钻孔	钻头 D4.5
底面	8	加工中心	加工 φ8 沉头孔	钻孔	D8
底面	9	加工中心	加工 φ13 的孔	钻孔	钻头 D13
上面	10	加工中心	粗加工	型腔铣	D16R0.4
上面	11	加工中心	粗加工	型腔铣	D10
上面	12	加工中心	二次开粗	剩余铣	D4
上面	13	加工中心	底面精加工	底壁铣	D10
上面	14	加工中心	精加工侧壁	底壁铣	D6
上面	15	加工中心	清角加工	带 IPW 的底壁铣	D4

1.2 打开模型文件进入加工模块

打开随书配套光盘，在例题文件夹中打开模型文件，并且进入加工模块。

(1) 启动 NX 12.0，单击左上角 【打开】按钮，在【打开】对话框中选择光盘"例题"文件夹中 1.prt 文件，单击【OK】按钮，如图 1-6 所示。

建立坐标系型腔铣

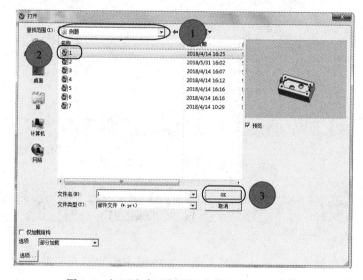

图 1-6　打开光盘"例题"文件夹中 1.prt 文件

(2) 单击【应用模块】按钮，再单击【加工】图标(也可以直接使用键盘快捷键 Ctrl+Alt+M)启动 NX 12.0"加工"模块，如图 1-7 所示。

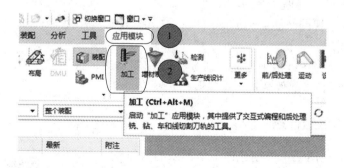

图 1-7　启动 UG NX 12.0"加工"模块

(3) 打开【加工环境】对话框，默认选择【CAM 会话配置】中【cam_general】，在【要创建的 CAM 组装】选项中选择【mill_planar】平面铣，单击【确定】按钮进入平面铣加工界面，如图 1-8 所示。

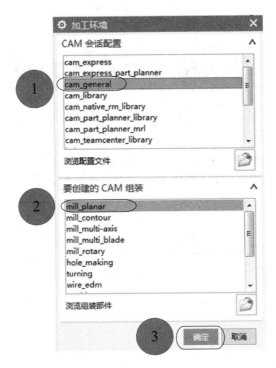

图 1-8　设置【加工环境】对话框

1.3　建立父节点组

父节点组包括几何视图、机床视图、程序顺序视图和加工方法视图。

(1) 几何视图：定义"加工坐标系"方向和安全平面，并设置"部件""毛坯"和"检查"几何体等参数。

(2) 机床视图：定义切削刀具。可以指定铣刀、钻头和车刀等，并保存与刀具相关的数据，以用作相应后处理命令的默认值。

(3) 程序顺序视图：能够把编好的程序按组排列在文件夹中，并按照从上到下的先后顺序排列加工程序。

(4) 加工方法视图：用来定义切削方法类型(粗加工、精加工、半精加工)。例如，"内公差""外公差"和"部件余量"等参数在此设置。

1.3.1　创建加工坐标系

在【几何】视图菜单中创建加工坐标系的操作步骤如下：

(1) 将【工序导航器】切换到【几何】视图页面，如图 1-9 所示。

(2) 双击【MCS_MILL】图标，弹出【MCS 铣

设置加工坐标系

削】对话框，如图 1-10 所示。

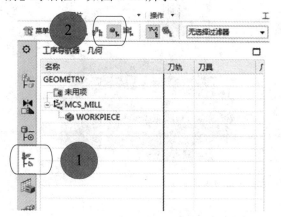

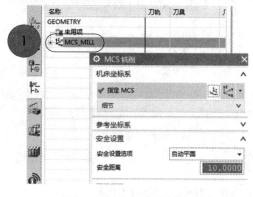

图 1-9 【几何】视图切换　　　　　　图 1-10 【MCS 铣削】对话框

(3) 由于毛坯材料没有经过机加工处理，表面形状可能不规则，因此为保证各面加工留量均匀，把坐标系放在毛坯中间位置。单击 【坐标系对话框】图标，如图 1-11 中圆圈 1 所示。在弹出的【坐标系】对话框中，选择【对象的坐标系】选项，如图 1-12 圆圈 2 所示。其功能是自动设置坐标为所选择平面的中心点位置。单击工件上表面方框自动捕捉工件的上表面中心点坐标位置，如图 1-13 圆圈 3 所示。在安全平面没有干涉物的情况下可以选择默认状态【自动平面】，安全距离为"10 mm"。如有干涉物，则可把安全距离设置为"50 mm～100 mm"，最后单击【确定】按钮退出【MCS 铣削】对话框，如图 1-11 中圆圈 4、5 所示。

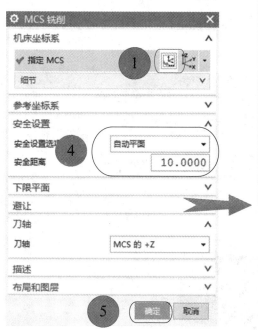

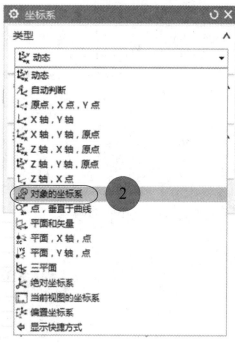

图 1-11 选择【坐标系对话框】图标　　　　图 1-12 选择【对象的坐标系】选项

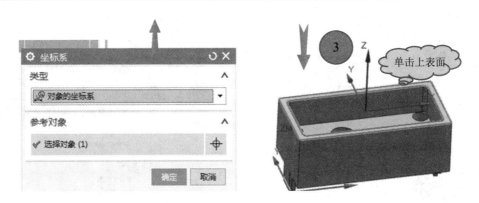

图 1-13　单击工件上表面方框设置坐标系

1.3.2　创建部件几何体

在【几何】视图菜单中创建加工部件几何体、零件毛坯以及检查几何体的操作步骤如下：

(1) 双击 WORKPIECE 图标，打开【工件】对话框，如图 1-14 所示。

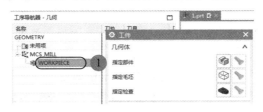

图 1-14　打开【工件】对话框

(2) 单击 【选择或编辑部件几何体】图标，如图 1-15 圆圈 1 所示，弹出【部件几何体】对话框。单击被加工工件，使其成橘黄色。然后单击【确定】按钮退出【部件几何体】对话框，如图 1-15 圆圈 2 所示。

图 1-15　创建部件几何体

(3) 单击 ⊞【选择或编辑毛坯几何体】图标，如图 1-15 方框 1 所示。弹出【毛坯几何体】对话框，在【类型】选项中选择【包容块】选项，如图 1-16 方框 2 所示。设置毛坯尺寸单边增加 "2.5 mm"，毛坯底面增加 "5 mm"，毛坯顶面不增加尺寸，如方框 3 所示，从而使毛坯尺寸达到 85 mm × 45 mm × 45 mm。单击【确定】按钮，如图 1-16 方框 4 所示，返回【工件】对话框，然后再单击【确定】按钮退出【工件】对话框完成设置，如图 1-16 方框 5 所示。

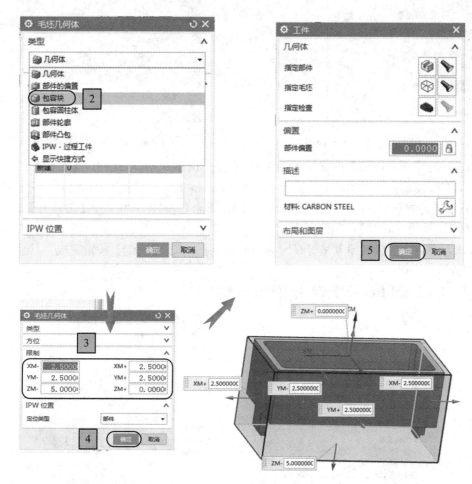

图 1-16　创建毛坯几何体

1.3.3　创建刀具

在【机床】视图下创建粗加工刀具 D16R0.4 钻铣刀的步骤如下：

(1) 单击 【机床】视图图标，如图 1-17 圆圈 1 所示，将【工序导航器】切换到【机床】视图页面。

(2) 单击 【创建刀具】图标，如图 1-18 圆圈 2 所示，弹出【创建刀具】对话框，如图 1-18 所示。

设置(选择)刀具

图 1-17　【机床】视图　　　　　　　　　　图 1-18　【创建刀具】对话框

(3) 选择 【平底刀】图标，如图 1-19 圆圈 1 所示，创建平底刀。【名称】位置输入 "D16R0.4"(代表直径为 16 mm，圆角半径为 0.4 mm 的镶片钻铣刀)，如图 1-19 所示。

(4) 平底刀参数设置中，直径设置为 "16"，下半径设置为 "0.4"，【刀具号】【补偿寄存器】【刀具补偿寄存器】三项均设置为 "1"(此数值代表刀具、刀具半径补偿和刀具长度补偿号，为避免发生撞机问题最好设置为相同数字)。单击【确定】按钮完成刀具建立，如图 1-20 所示。

其他刀具在编辑加工程序前，按照给定参数自行设置。

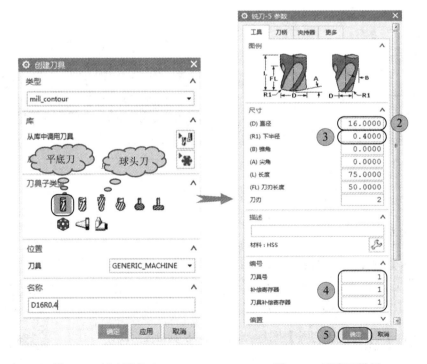

图 1-19　创建平底刀　　　　　　　　　　图 1-20　平底刀设置

1.3.4　创建程序组

在【程序顺序】视图中创建加工程序组文件夹，操作步骤如下：

(1) 将【工序导航器】切换到【程序顺序】视图页面，如图 1-21 所示。

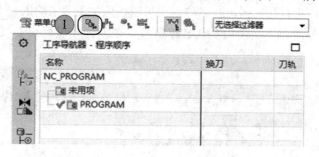

图 1-21　【工序导航器】切换到【程序顺序】视图页面

(2) 双击【PROGRAM】程序组文字，修改文件名为"底面"(使用右键单击【PROGRAM】程序组，选择【重命名】也可实现更改名称)，如图 1-22 圆圈 2 所示。

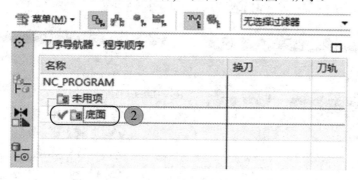

图 1-22　双击修改程序组名称

(3) 保存文件。

1.4　开关盒中框的第一序加工

由前面的加工分析得出此工件的第一序加工(粗加工)是为零件底部朝上加工内外型腔，留出上面两凸起部分，再进行二序加工。以下为粗加工程序编制方法。

1.4.1　粗加工程序编制

NX 软件在建立有模型图的基础上编程时，最为简单有效的开粗程序就是曲面加工里的型腔铣，使用型腔铣可以完成绝大多数零件的开粗工作。接下来学习怎样使用型腔铣的加工方法完成零件粗加工程序的编制。

为了更好地体现出加工程序的先后顺序，本案例全部使用【程序顺序】视图来完成程序的编制。

(1) 单击【创建工序】图标，如图 1-23 圆圈 1 所示，弹出【创建工序】对话框。

(2) 在【类型】下拉菜单中选择【mill_contour】曲面铣选项，如图 1-24 圆圈 2 所示。选择 【型腔铣】图标(见图 1-25 圆圈 3)，在【程序】下拉菜单中选择刚建好的【底面】程序组，【刀具】下拉菜单中选择【D16R0.4】的镶片钻铣刀，【几何体】下拉菜单中选择建立好的【WORKPIECE】几何体，在名称栏中可以按照加工要求输入一个程序名称(本书中不对程序名做专门修改，直接使用默认名称填写)，然后单击【确定】按钮，如图 1-25 圆圈 4、5 所示。

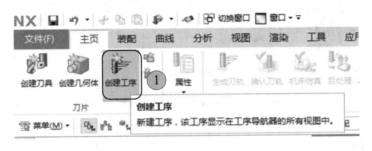

图 1-23 单击【创建工序】图标

图 1-24 选择【mill_contour】曲面铣选项

图 1-25 创建【型腔铣】工序

(3) 在弹出的【型腔铣】对话框中，只要在【几何】视图中正确设定【WORKPIECE】，那么在进入型腔铣时，【指定部件】和【指定毛坯】选项应显示为灰色，当右侧 【显示】图标为彩色时，单击可显示已选择的几何体部分。如果进入型腔铣后还能选择【指定部件】和【指定毛坯】，则说明【几何】视图中的【WORKPIECE】没有设定，或进入程序前没有选择几何体为【WORKPIECE】。型腔铣编程中一般不用设置【指定切削区域】，【指定检查】和【指定修剪边界】根据零件加工需求设置，在本例中不需要设置，如图 1-26 所示。

(4) 打开【工具】菜单下拉箭头，显示出已选择的加工【刀具】为【D16R0.4】镶片钻铣刀(注：此项工作前序选择正确的情况下可忽略)，如图 1-27 所示。

(5) 打开【刀轴】菜单下拉箭头显示出默认刀轴为【+ZM 轴】，三轴加工中心刀轴一般使用+ZM 轴，只有在使用多轴机床加工时才会修改此项(注：此选项三轴加工编程时不用选择，使用默认设置即可)，如图 1-27 所示。

图 1-26　几何体设置对话框　　　　　　　　图 1-27　工具和刀轴选项

(6)【刀轨设置】为型腔铣参数设置的主要内容。【切削模式】选项中一般常用 跟随部件 和 跟随周边 两种，【跟随部件】适合加工开放轮廓的工件，可以使刀具从外向内加工，并从工件外下刀。【跟随周边】更适合加工封闭轮廓的工件，可以使刀具从内向外加工，减少型腔加工时的下刀位置变化。此案例底部封闭轮廓加工面较多，所以在【切削模式】中选择【跟随周边】的加工方法，如图 1-28 圆圈 1 所示。

(7) D16R0.4 钻铣刀粗加工时，XY 方向的刀具步距一般使用刀具直径的 70%～80%，精加工时步距使用刀具直径的 50%以下，80%～100%的 XY 方向步距一般情况下不推荐使用。首先，若刀具步距太大，则每次切削时相当于满刀切削，刀具受力过大会影响刀具寿命和机床精度；其次，步距太大的话，加工完底面，光洁度很低且会出现接刀痕。因此在【平面直径百分比】选项中设置数值为"75"，如图 1-29 圆圈 2 所示。

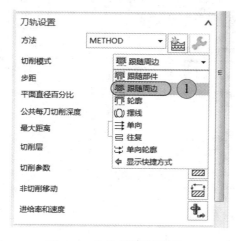
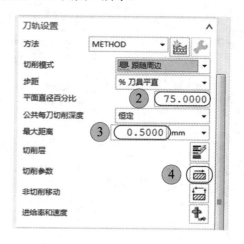

图 1-28　选择【跟随周边】　　　　　　　　图 1-29　XYZ 方向步距量设置

(8) D16R0.4 钻铣刀粗加工时的 Z 方向步距一般情况下取值为每层 0.3 mm～0.7 mm。【最大距离】选项设置的就是刀具 Z 方向的每层步距量，因此设置中间数值每层"0.5 mm"，如图 1-29 圆圈 3 所示。

(9) 单击 ⚏【切削参数】图标，如图 1-29 圆圈 4 所示，打开【切削参数】对话框。

(10) 在【策略】标签中设置【切削顺序】为【深度优先】,【深度优先】会按照不同区域分别由上往下加工，以减少抬刀和过刀路径，减少加工时间。【层优先】会按照同一深度在不同区域跳刀加工，从而增加很多抬刀路径，增加加工时间。一般情况下优先选择【深度优先】，如图 1-30 所示。

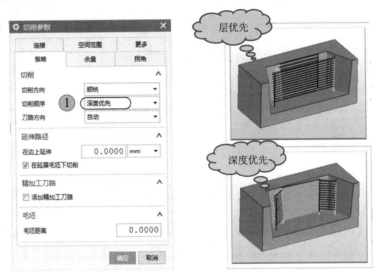

图 1-30　【策略】标签设置

(11) 选择【余量】标签，设置【部件侧面余量】参数为"0.2 mm"(注：粗加工刀具余量一般设置为"0.2 mm")，单击【确定】按钮退出【切削参数】对话框，如图 1-31 所示。

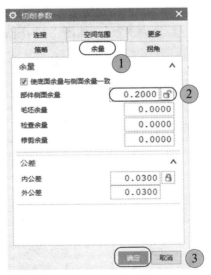

图 1-31　【余量】标签设置

(12) 单击 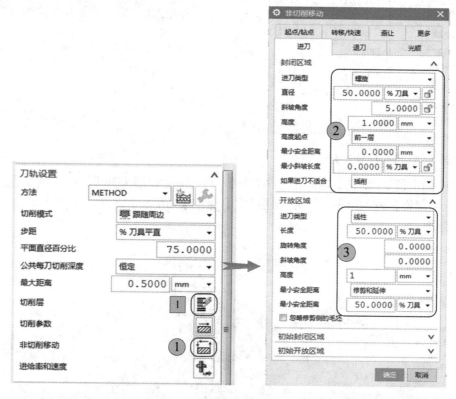【非切削移动】图标，如图 1-32 圆圈 1 所示，打开【非切削移动】对话框，如图 1-33 所示。

(13) 设置进刀参数，首先设置【封闭区域】下刀参数，型腔内下刀一般采用【螺旋】下刀的方式。按照螺旋下刀加工原理可知，在型腔尺寸能够满足螺旋下刀时，选用螺旋下刀的方式进刀。在型腔尺寸不能满足螺旋下刀时，选用斜线下刀或直线下刀的方式进行。螺旋下刀【直径】选用刀具直径的"50%"，【斜坡角度】设置为"5°"，【高度】设置为"1 mm"，【最小斜坡长度】设置为"0"(注：加工刀具能够直线下刀时可以输入数字为 0。如果加工刀具不能直线下刀，那么此数字最小不能小于 50，否则会发生撞机事故)，如图 1-33 圆圈 2 所示。【开放区域】设置进刀【长度】为刀具的"50%"，抬刀【高度】设置为"1 mm"，以减少抬刀距离。具体参数设置如图 1-33 圆圈 3 所示。

图 1-32 选择非切削移动　　　　图 1-33 进刀参数设置

(14) 单击【非切削移动】对话框【转移/快速】标签设置快速抬刀高度。为提高加工速度，应尽量减少抬刀高度，所以把【区域之间】和【区域内】的【转移类型】都改为【前一平面】，并且把抬刀【安全距离】都设置为"1 mm"，如图 1-34 圆圈 1 所示。这样做的好处是能减少不必要的抬刀，节约加工时间。但大部分快速移刀都是在工件零平面以下进行，所以要求机床的 G00 运动必须是两点间的直线运动，不能是两点间的折线运动，否则会发生撞机事故。加工前一定要在 MDI 下输入"G00 走斜线观察机床"的移动方式，如果不对，则需要修改机床参数。或者按照 NX12.0【转移/快速】的初始设置方法，把【转移类型】全部设置为【安全距离-刀轴】。最后，单击【确定】按钮退出【非切削移动】对话框，具体参数设置如图 1-34 所示。

(15) 单击 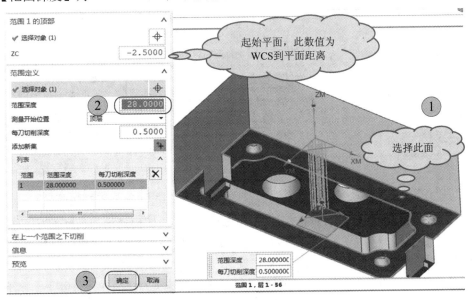【切削层】图标，如图 1-32 方框 1 所示，进入【切削层】对话框。

(16) 在【切削层】对话框中单击 ✖ 多次将列表中数值全部删除，如图 1-35 所示。

图 1-34 【转移/快速】标签设置

图 1-35 删除列表数值

(17) 单击【范围定义】里的【选择对象】，选择零件如图 1-36 圆圈 1 所示位置。测得加工【范围深度】为 "28 mm"，单击【确定】退出【切削层】对话框。

图 1-36 选择切削深度

(18) 单击 ✛【进给率和速度】图标，打开【进给率和速度】对话框，设置【主轴速度】为 "2500 rpm"(注：输入 "2500" 后一定要点击后面的【计算器】图标，否则会报警)。【进给率】中的【切削】设置为 "1500 mmpm"，输入【进刀】为 "70%切削"。然后，单击【确定】按钮退出【进给率和速度】对话框，如图 1-37 所示。

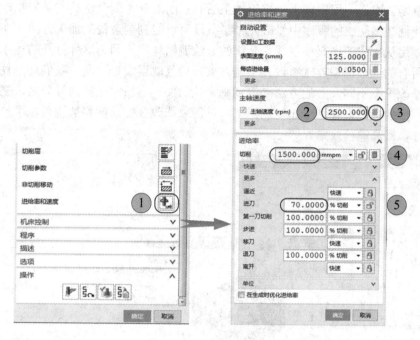

图 1-37　进给率和速度设置

（19）单击 ![生成图标] 【生成】图标，如图 1-38 圆圈 1 所示，计算加工路径。单击【确定】退出型腔铣，如图 1-38 圆圈 2 所示。

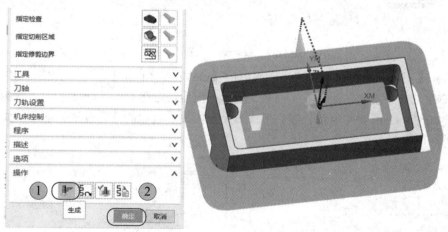

图 1-38　生成刀具路径

1.4.2　二次开粗程序编制

因为型腔内凹圆角的半径为 4 mm，所以选用直径为 6 mm 的平底刀进行圆角残料的二次开粗工作。

（1）点开【机床】视图，然后建立一把直径为 6 mm 的硬质合金刀，刀号为 2 号，刀具名称为 D6。

(2) 二次开粗程序在 NX 12.0 软件中所使用的名称为"剩余铣",如图 1-39 所示。剩余铣的作用就是去除上一程序或上一把刀加工后剩余的部分。剩余铣其实是属于型腔铣的一个分支,我们在型腔铣的程序中修改几个参数就可以得到剩余铣的加工方法。在实际应用中一般不使用从创建工序直接创建剩余铣的方法编辑程序,因为这样打开的程序在初始化界面里,所有的选项都需要重新设置一遍,太繁琐且耽误编程时间。我们采用复制上一程序,然后修改里面不同参数的方法来得到剩余铣的加工方法,这样的好处是,复制下来的程序中原程序的参数设置都一起复制下来,只需要修改有变动的参数就可以了,能够减少编程的工作,提高编程效率。

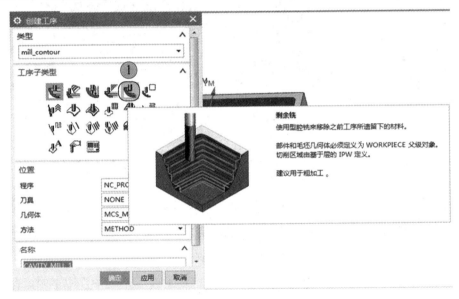

图 1-39　剩余铣加工方法选择

(3) 在【机床】视图中右键单击刚编好的【CAVITY_MILL】程序,如图 1-40 圆圈 1 所示,右键选择【复制】选项,如圆圈 2 所示。然后,右键单击【D6】刀具,选择【内部粘贴】,如图 1-41 圆圈 4 所示。最后形成一个新的提示错误的【型腔铣】程序,如图 1-42 圆圈 5 所示。

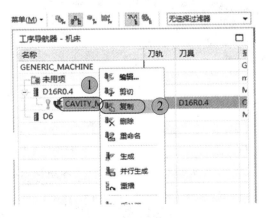

图 1-40　复制【型腔铣】程序

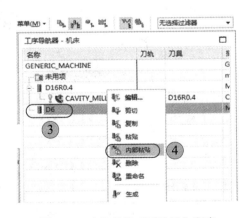

图 1-41　内部粘贴【型腔铣】程序

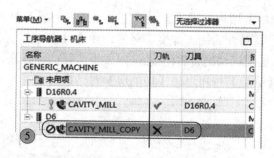

图 1-42　新的【型腔铣】程序

（4）双击打开新复制过来的【型腔铣】程序。设置 Z 向分层【最大距离】为"0.2 mm"，如图 1-43 所示。

（5）单击【切削参数】图标，如图 1-44 圆圈 1 所示，修改【切削参数】设置。

图 1-43　设置 Z 向分层为"0.2 mm"每刀　　　图 1-44　单击【切削参数】图标

（6）在【切削参数】对话框中【空间范围】选项中设置工件剩余毛坯。在【过程工件】下拉菜单中选择【使用基于层的】选项，设置本次加工毛坯为上次加工剩余的部分。在【剩余铣】中此选项默认为【使用基于层的】选项，【型腔铣】和【剩余铣】的不同只有这一个选项，然后单击【确定】按钮，如图 1-45 所示。

图 1-45　设置【空间范围】标签

(7) 单击▤【切削层】图标，设置加工深度。在【切削层】对话框中单击【列表】右侧✕【移除】图标，删除原有的加工深度。然后，单击【范围定义】里的【选择对象】，选择工件型腔底面作为加工最深的位置(注意：如果不选择此项程序，则会加工中间圆孔)。如图 1-46 所示，单击【确定】按钮退出【切削层】对话框。

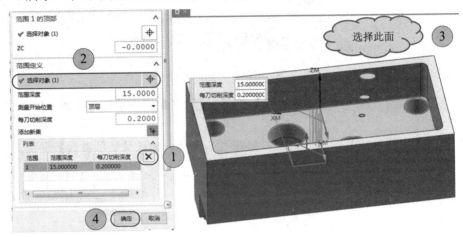

图 1-46　设置加工深度

(8) 单击✦【进给率和速度】图标，设置【主轴速度】为"4000 rpm"，单击转速右侧▤【计算器】图标。最后，单击【确定】按钮退出【进给率和速度】对话框，如图 1-47 所示。

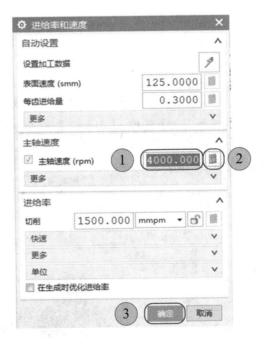

图 1-47　设置【主轴速度】

(9) 单击▶【生成】图标，如图 1-48 圆圈 1 所示。计算加工程序，生成【剩余铣】加工路径，单击【确定】按钮退出【剩余铣】设置，如图 1-48 圆圈 2 所示。

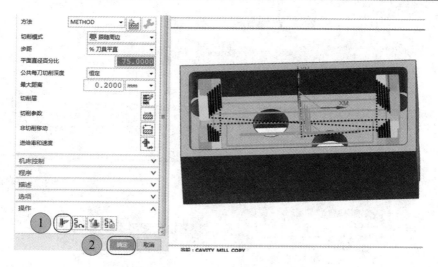

图 1-48   生成【剩余铣】加工路径

1.4.3 精加工底面程序编制

使用 φ10 硬质合金刀精加工型腔底面程序的编制方法如下：

(1) 点开【机床】视图，然后建立一把直径为 10 mm 的硬质合金刀，刀号为 3 号，刀具名称为 D10。

凹腔底面精加工程序

(2) 创建精加工底面程序，单击 【创建工序】图标，在【类型】下拉菜单中选择【mill_planar】平面铣选项，在【工序子类型】选项中选择 ⊔【底壁铣】加工方法，【程序】选项中选择【底面】程序组，【刀具】选项中选择刚创建的【D10】铣刀，【几何体】选项中选择【WORKPIECE】，最后，单击【确定】按钮进入【底壁铣】对话框，如图 1-49 所示。

图 1-49   进入【底壁铣】加工

(3) 单击【指定切削区域底面】图标选择要加工的底面。单击型腔底面，如图 1-50 圆圈 2 所示，单击【确定】退出【切削区域】对话框，如图 1-50 圆圈 3 所示。

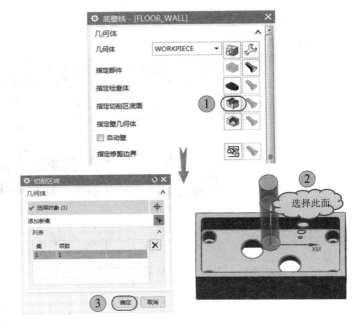

图 1-50　【切削区域】对话框选择底面

(4) 勾选 ☐ 自动壁 【自动壁】图标前的方框，使其自动捕捉出和已选择底面相邻的工件侧壁为壁几何体。

(5) 因为被加工表面为凹平面，所以更适合使用【跟随周边】的加工方法。设置【切削模式】为【跟随周边】，如图 1-51 圆圈 1 所示。精加工时 XY 方向的步距量应小于刀具直径的 50%。软件中默认为 50%，故不需要修改。【底面毛坯厚度】设置为粗加工时底面的余量，在此选项框中输入"0.2 mm"，如图 1-51 圆圈 2 所示。

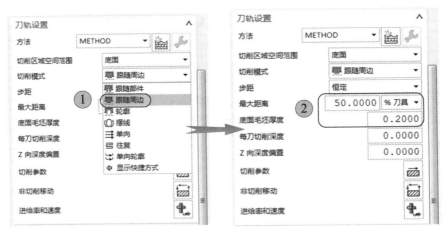

图 1-51　【刀轨设置】对话框设置

(6) 单击 🔳 【切削参数】图标，打开【切削参数】对话框，单击上方【余量】标签，设置【壁余量】为"0.1 mm"(注意：由于此程序是精加工底面程序，所以【最终底面余量】应为"0 mm"。工件壁的精加工应由专门的程序执行，所以在【壁余量】选项中输入"0.1 mm"，如图 1-52 圆圈 2 所示，给精加工壁的程序留有单边 0.1 mm 的余量)。减小公差的数值提高加工精度，设置【内公差】和【外公差】为"0.01 mm"，单击【确定】按钮退出【切削参数】对话框，如图 1-52 所示。

（7）在【非切削移动】对话框中单击左上方【进刀】标签，在【进刀】标签中设置进刀参数，如图 1-53 所示。

（8）单击 【进给率和速度】图标，设置【主轴速度】为"3500 rpm"，单击右侧 【计算器】图标。设置【进给率】【切削】为"1000 mmpm"。设置【更多】中【进刀】选项为"70%"以降低进刀速度，如图 1-54 圆圈 3 所示，然后单击【确定】按钮退出【进给率和速度】对话框，如图 1-54 所示。

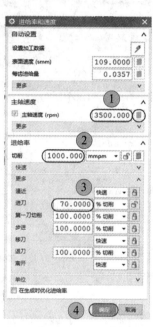

图 1-52　加工余量设置　　　　图 1-53　进刀参数设置　　　图 1-54　【进给率和速度】对话框

（9）单击 【生成】图标，如图 1-55 圆圈 1 所示，计算加工程序。生成【底壁铣】加工路径，单击【确定】按钮退出【底壁铣】设置，如图 1-55 圆圈 2 所示。

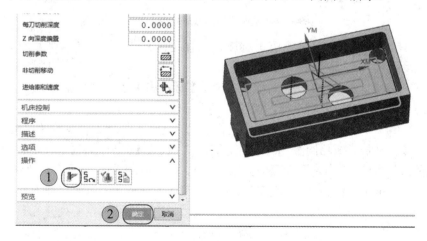

图 1-55　生成底壁铣加工路径

1.4.4 精加工侧面程序编制

使用 φ6 和 φ10 合金刀精加工零件内外轮廓程序的编制方法如下：

(1) 把精加工底面程序修改为精加工侧壁的程序。首先，在【机床】视图中，右键单击刚编好的【底壁铣】程序，如图 1-58 圆圈 2 所示，选择【复制】选项，如图 1-56 圆圈 3 所示，然后右键单击【D6】刀具，选择【内部粘贴】，如图 1-57 圆圈 4、5 所示。最后，形成一个新的提示错误的【底壁铣】程序，如图 1-58 圆圈 6 所示。

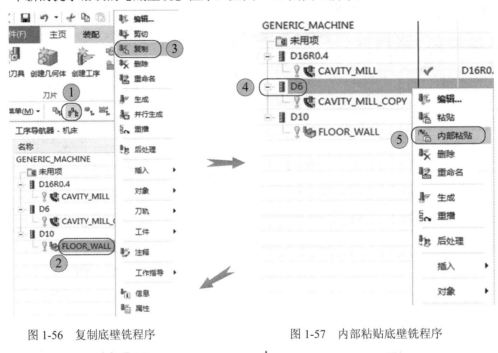

图 1-56 复制底壁铣程序　　　　　　　　图 1-57 内部粘贴底壁铣程序

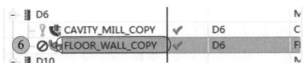

图 1-58 复制出新的【底壁铣】程序

(2) 双击打开新复制的【底壁铣】程序。设置【切削模式】为【轮廓】的加工方法，如图 1-59 所示。

(3) 单击 ⊞ 【切削参数】图标，设置【余量】标签中【壁余量】为"0 mm"。单击【确定】按钮退出【切削参数】对话框，如图 1-60 所示。

(4) 单击 ⊞ 【非切削移动】图标，选择左上方【进刀】标签，设置【封闭区域】【进刀类型】为【与开放区域相同】，如图 1-61 圆圈 2 所示。【开放区域】【进刀类型】设置为【圆弧】的进刀方式，【半径】设置为刀具直径的"50%"，【圆弧角度】为"90°"，进刀【高度】为"1 mm"，【最小安全距离】选择【无】。由于精加工需要打开刀具半径补偿功能，G41 和 G42 刀补必须建立在直线路径上，所以勾选【在圆弧中心处开始】选项，建立进刀圆弧的直线延长线。如果不打开此项，则在设置刀具半径补偿时，没有直线路径刀补会建立在圆弧上，机床加工中会产生警报，具体参数设置如图 1-61 所示。

图 1-59　选择【轮廓】加工方法

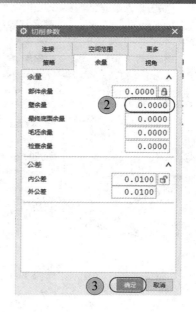

图 1-60　【壁余量】设置为"0"

(5) 单击【更多】标签，如图 1-62 圆圈 5 所示，打开刀具半径补偿功能，设置【刀具补偿位置】为【所有精加工刀路】，取消【最小移动】和【最小角度】中的数值，全部改写为"0 mm"，如图 1-62 圆圈 6 所示，然后单击【确定】按钮退出【非切削移动】对话框。

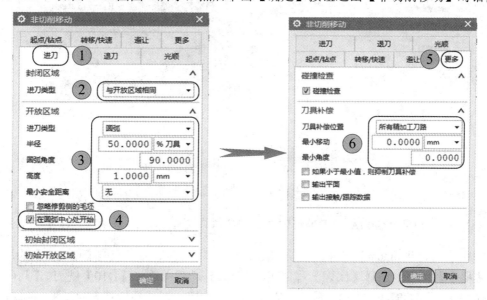

图 1-61　进刀圆弧设置

图 1-62　打开刀具半径补偿功能

(6) 单击 ![进给率和速度图标] 【进给率和速度】图标，打开【进给率和速度】对话框，修改【进给率】【切削】为"500 mmpm"(精加工侧壁时走刀速度不宜过快，应控制在 500 mmpm 以下。否则，加工表面粗糙度过大会影响加工质量)。然后，单击【确定】按钮退出【进给率和速度】对话框，如图 1-63 所示。

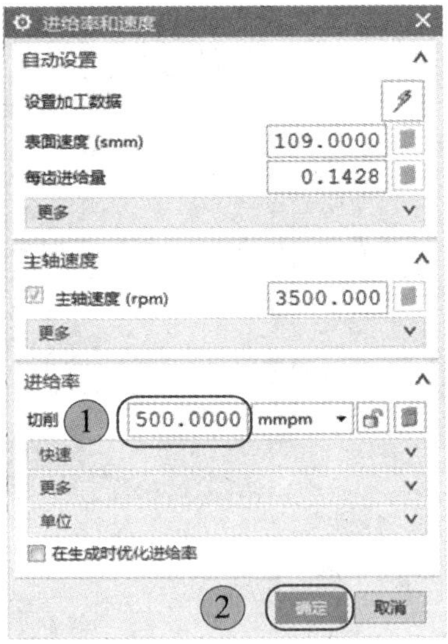

图 1-63　设置进给速度

(7)　单击【生成】图标，如图 1-64 圆圈 1 所示，计算加工程序，生成【底壁铣】【精加工侧壁】路径，单击【确定】按钮退出【底壁铣】设置，如图 1-64 圆圈 2 所示。

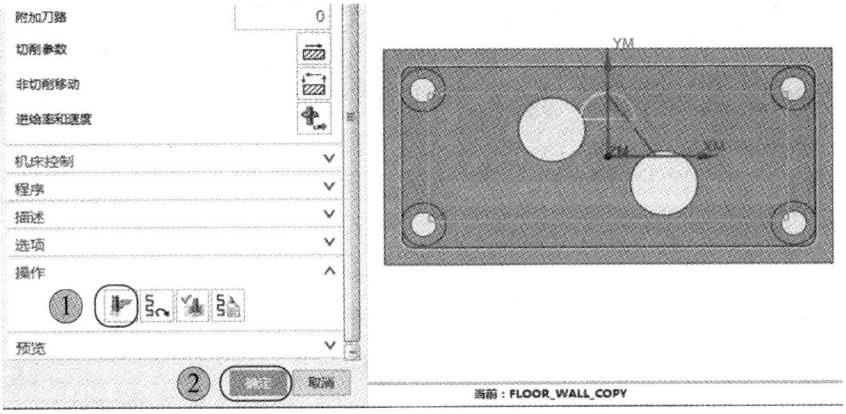

图 1-64　生成【精加工侧壁】程序

(8) 使用φ10 合金刀精加工外形轮廓。单击 创建工序 【创建工序】图标，在【类型】下拉菜单中选择【mill_planar】平面铣选项，【工序子类型】选项中选择【精铣壁】的加工方法，【程序】选项中选择【底面】程序组，【刀具】选项中选择【D10】铣刀，【几何体】选项中选择【WORKPIECE】，最后单击【确定】按钮进入【精铣壁】对话框，如图 1-65 所示。

(9)　单击【指定部件边界】图标，如图 1-66 圆圈 1 所示，打开【部件边界】对话框，在【边界】的【选择方法】选项中选择【曲线】的方式创建加工边界，如图 1-66 圆圈 2 所示。设置【边界类型】为【封闭】，【刀具侧】为【外侧】，【平面】为【自动】，如图 1-66 圆圈 3 所示。然后，手动选择轮廓外圈线为加工边界，单击【确定】退出设置，如图 1-66 圆圈 5、6 所示。

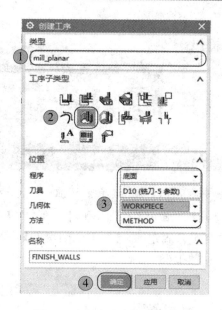

图 1-65 创建【精铣壁】程序

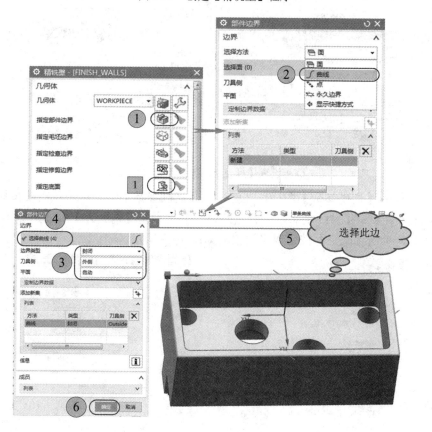

图 1-66 设置【部件边界】

(10) 单击【指定底面】图标,如图 1-66 方框 1 所示,弹出【平面】对话框。选择底部平面为加工底面,如图 1-67 所示。

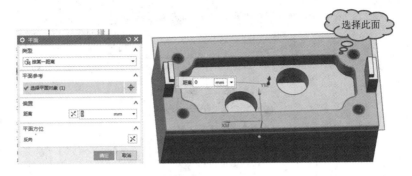

图 1-67　选择加工底面

(11) 单击 ▤ 【切削层】图标，弹出【切削层】对话框，在【类型】下拉菜单中选择【恒定】的切削方式，即每层的高度是固定的，【每刀切削深度】【公共】设置为"14 mm"，单击【确定】按钮退出设置，如图 1-68 所示。

图 1-68　设置【每刀切削深度】为"14 mm"

(12) 单击 ▦ 【切削参数】图标，在弹出的【切削参数】对话框中单击【余量】标签，设置【部件余量】和【最终底面余量】全部为"0"，设置【内公差】和【外公差】数值全部为"0.01"，单击【确定】按钮退出设置，如图 1-69 所示。

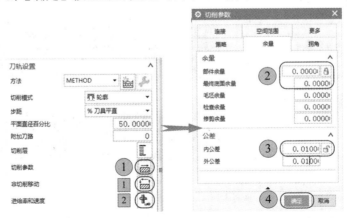

图 1-69　设置加工余量和内外公差

(13) 单击🔲【非切削移动】图标，如图 1-69 方框 1 所示，打开【非切削移动】对话框，设置【进刀类型】为【圆弧】，并且打开半径补偿功能，具体设置方法和【底壁铣】方法相同，如图 1-70 所示。

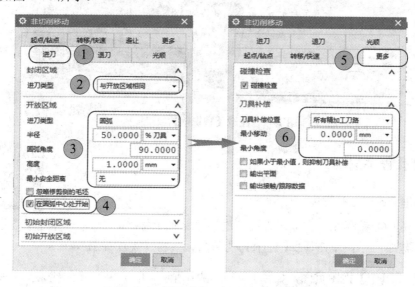

图 1-70　设置【非切削移动】对话框

(14) 单击🔩【进给率和速度】图标，如图 1-69 方框 2 所示，打开【进给率和速度】对话框，修改【进给率】【切削】为"500 mmpm"(精加工侧壁时走刀速度不宜过快，应控制在 500 mmpm 以下。否则，加工表面粗糙度过大会影响加工质量)。然后，单击【确定】按钮，退出【进给率和速度】对话框，如图 1-71 所示。

图 1-71　设置主轴转速和进给速度

(15) 单击▶【生成】图标，如图 1-72 圆圈 1 所示。计算加工程序，生成【精铣壁】【精加工侧壁】路径，单击【确定】按钮退出【精铣壁】设置，如图 1-72 圆圈 2 所示。

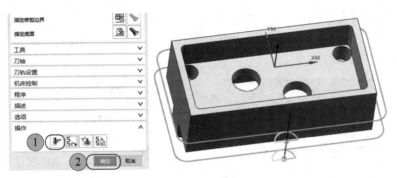

图 1-72　生成【精铣壁】加工程序

1.4.5　打孔程序编制

编辑 6 个孔位加工程序

1. 创建 φ6 中心钻程序

创建 φ6 中心钻编制，完成工件孔位的中心钻加工程序。

(1) 创建中心钻，单击 ![]【机床】视图图标，将【工序导航器】切换到【机床】视图页面，如图 1-73 圆圈 1 所示。单击 ![]【创建刀具】图标，如图 1-73 圆圈 2 所示，弹出【创建刀具】对话框。

(2) 在【类型】下拉菜单中选择【hole_making】孔加工类型。在【刀具子类型】对话框中选择 ![]【中心钻】图标，在【名称】对话框中输入刀具名称"中心钻 D6"，然后单击【确定】按钮，如图 1-73 所示，弹出【中心钻刀】对话框。

图 1-73　【创建刀具】对话框

(3) 输入刀具【直径】为"6 mm"，【刀具号】和【补偿寄存器】都输入"4"，其他参数无需设置，按默认值就可以。最后，单击【确定】按钮退出设置，如图 1-74 所示。

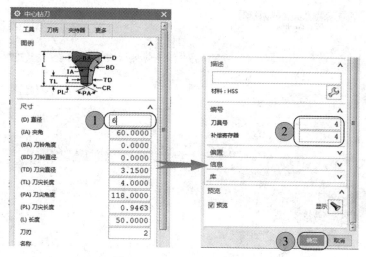

图 1-74 【中心钻刀】设置

(4) 创建点中心孔程序，单击 【创建工序】图标，打开【创建工序】对话框，在【类型】下拉菜单中选择【hole_making】孔加工选项，【工序子类型】选项中选择 【定心钻】加工方法，如图 1-75 圆圈 2 所示。【程序】选项中选择【底面】程序组，【刀具】选项中选择刚创建的【中心钻 D6】刀具，【几何体】选项中选择【WORKPIECE】，如图 1-75 圆圈 3 所示。最后，单击【确定】按钮打开【定心钻】对话框。

(5) 在【定心钻】对话框中单击 【指定特征几何体】图标，如图 1-76 圆圈 5 所示。选择中心钻的六个孔位，如图 1-77 圆圈 1 所示，然后按住键盘按键【Shift】。单击列表中最下面的一个孔和最上面的一个孔，松开【Shift】按键，这时列表中的所有孔都显示为蓝色背景状态。单击 图标，选择 用户定义(U) 选项，【深度】数值输入"1 mm"为点孔深度。最后，单击【确定】退出设置，如图 1-77 所示。

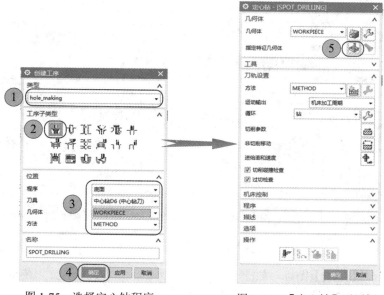

图 1-75 选择定心钻程序　　　　　图 1-76 【定心钻】对话框

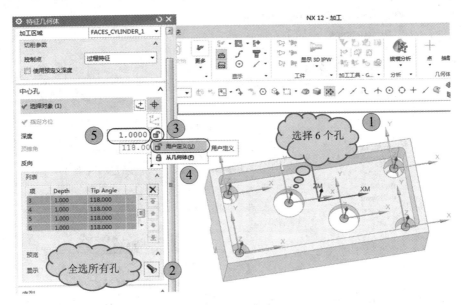

图 1-77　选择孔位设置加工深度

（6）单击 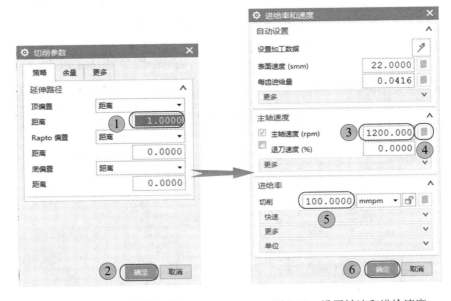 【切削参数】图标，打开【切削参数】对话框，设置【延伸路径】【距离】为"1 mm"，减小打孔起始距离，提高加工速度，如图 1-78 所示，然后单击【确定】退出对话框。

（7）单击 【进给率和速度】图标打开【进给率和速度】对话框，设置【主轴速度】为"1200 rpm"，单击转速右侧【计算器】图标，设置【进给率】【速度】为"100 mmpm"，然后单击【确定】速度退出对话框，如图 1-79 所示。

图 1-78　设置打孔起始距离　　　图 1-79　设置转速和进给速度

（8）单击 【生成】图标，如图 1-80 圆圈 1 所示，计算加工程序，生成【定心钻】加工路径，单击【确定】按钮退出【定心钻】设置，如图 1-80 所示。

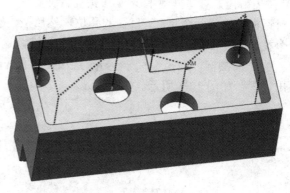

图 1-80　生成定心钻加工程序

2．创建四个 ϕ4.5 圆孔打孔程序

创建 ϕ4.5 钻头编辑完成零件上四个 ϕ4.5 圆孔的打孔程序。

（1）在【机床】视图下，按照创建中心钻的方法建立一把 ϕ4.5 的钻头，刀具号为 5 号，刀具名称为【钻头 D4.5】，参考图 1-73 及图 1-74。

（2）创建钻孔程序，单击 📌【创建工序】图标，打开【创建工序】对话框，在【类型】下拉菜单中选择【hole_making】孔加工选项，【工序子类型】选项中选择 🔧【钻孔】加工方法，如图 1-81 圆圈 1 所示。【程序】选项中选择【底面】程序组，【刀具】选项中选择刚创建的【钻头 D4.5】刀具，【几何体】选项中选择【WORKPIECE】，如图 1-81 圆圈 2 所示，最后单击【确定】按钮，如图 1-81 圆圈 3 所示，打开【钻孔】对话框，如图 1-82 所示。

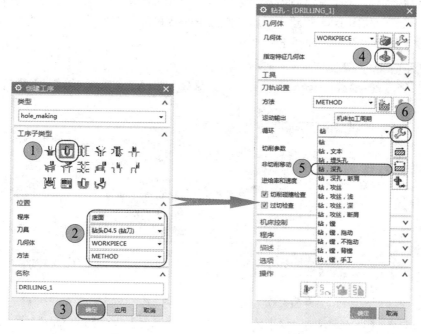

图 1-81　创建钻孔程序　　　　　　　图 1-82　选择孔位和啄钻方法

(3) 在【钻孔】对话框中单击 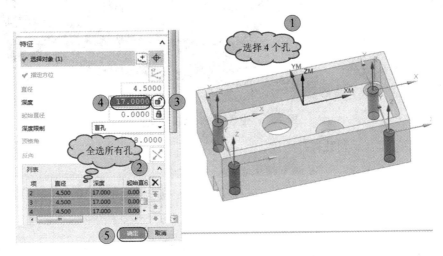【指定特征几何体】图标，如图 1-82 圆圈 4 所示。在弹出的对话框中选择四个 φ4.5 直径的孔位，然后按住键盘按键【Shift】。单击列表中最下面的一个孔和最上面的一个孔，松开【Shift】按键，这时列表中的所有孔都显示为蓝色背景状态。单击 🔒 图标，选择 🔓 用户定义(U) 选项，打孔【深度】设置为"17 mm"，单击【确定】退出设置，如图 1-83 所示。

图 1-83　选择四个孔位设置加工【深度】为"17 mm"

(4) 单击 【切削参数】图标，打开【切削参数】对话框，设置顶偏置距离为"1 mm"，以减小打孔起始距离，提高加工速度，然后单击【确定】退出对话框，如图 1-84 所示。

(5) 在【循环】下拉菜单下选择【钻，深孔】选项，设置为啄钻加工，如图 1-82 圆圈 5 所示。然后，单击 【编辑循环】图标，如图 1-82 圆圈 6 所示，设置【步进】【最大距离】为"1 mm"，如图 1-85 圆圈 3 所示，此数值是啄钻加工中每层打孔的深度值。最后，单击【确定】退出对话框，如图 1-85 所示。

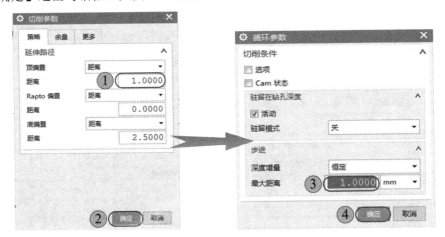

图 1-84　设置打孔起始距离　　　　图 1-85　设置啄钻每层的深度

(6) 单击 【进给率和速度】图标，设置【主轴速度】为"1000 rpm"，单击转速右侧【计算器】图标，设置【进给率】【切削】为"100 mmpm"，然后，单击【确定】退出对话框，如图 1-86 所示。

图 1-86　设置转速和进给速度

(7) 单击 ![生成]图标，如图 1-87 圆圈 1 所示计算加工程序，生成【钻孔】加工路径，单击【确定】按钮退出【钻孔】设置，如图 1-87 圆圈 2 所示。

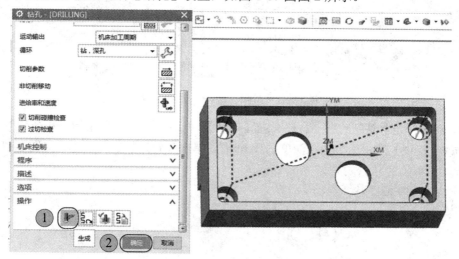

图 1-87　生成钻孔加工程序

3. 创建四个 φ8 深 10 mm 沉头孔程序

创建 φ8 平底刀，完成工件上四个 φ8 平头孔的加工程序。

(1) 在【机床】视图下，按照创建平底刀的方法建立一把 φ8 的平底刀，刀号为 6 号，刀具名称为【D8】，参考图 1-17～图 1-20。

(2) 把 φ4.5 钻头程序修改为 φ8 沉头孔的程序。首先，在【机床】视图中右键单击刚编好的【钻孔】程序，选择【复制】选项，如图 1-88 圆圈 1 和 2 所示。然后，右键单击【D8】

刀具，选择【内部粘贴】，如图 1-89 圆圈 3、4 所示，形成一个新的提示错误的【钻孔】程序，如图 1-90 所示。

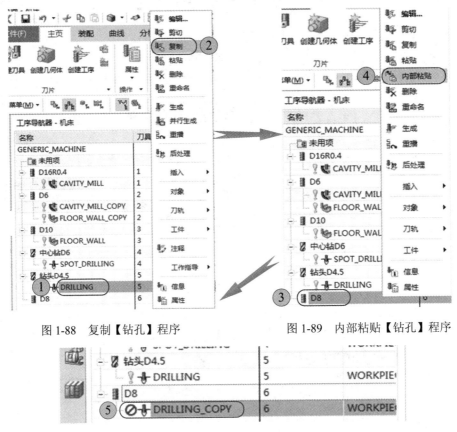

图 1-88　复制【钻孔】程序　　　　　　图 1-89　内部粘贴【钻孔】程序

图 1-90　复制出新的【钻孔】程序

(3) 双击打开新复制的钻孔程序。在【钻孔】对话框中单击 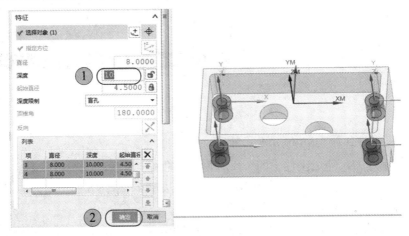 【指定特征几何体】图标，修改【深度】数值为沉头孔深度"10 mm"，单击【确定】退出设置，如图 1-91 所示。

图 1-91　修改打孔【深度】为"10 mm"

(4) 单击 【进给率和速度】图标，打开【进给率和速度】对话框，设置【主轴速度】为"2000 rpm"，单击转速右侧【计算器】图标，设置【进给率】【切削】为"60 mmpm"，然后，单击【确定】按钮退出对话框，如图 1-92 所示。

图 1-92　设置转速和进给速度

(5) 单击 【生成】图标计算加工程序，生成【钻孔】加工路径，单击【确定】按钮退出【钻孔】设置，如图 1-93 所示。

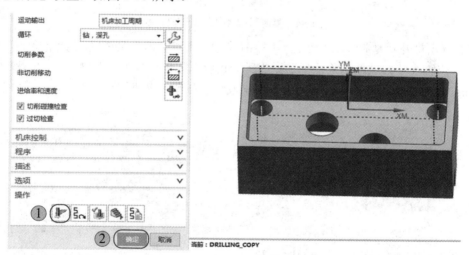

图 1-93　生成【钻孔】加工路径

4. 创建两个 φ13 通孔程序

创建 φ13 钻头，完成工件上两个 φ13 通孔的加工程序。

(1) 在【机床】视图下，按照创建钻头的方法建立一把 φ13 的钻头，刀号为 7 号，刀具名称为【钻头 D13】，参考图 1-73～图 1-74。

(2) 把 φ4.5 钻头程序修改为 φ13 通孔的程序，单击【机床】视图图标，如图 1-94 圆圈 1 所示，右键单击钻头 D4.5 的【DRILLING】钻孔程序，如图 1-94 圆圈 2 所示，选择【复制】选项，如图 1-94 圆圈 3 所示。然后，右键单击【钻头 D13】刀具，选择【内部粘贴】，如图 1-94 圆圈 4、5 所示，形成一个新的提示错误的【钻孔】程序，如图 1-94 圆圈 6 所示。

图 1-94　复制新的【钻孔】刀具路径

　　(3) 双击打开新复制过来的钻孔程序。在【钻孔】对话框中单击　【指定特征几何体】图标，单击列表中　图标删除原先选择的所有孔位，重新选择两个 φ13 的孔位。修改打孔孔【深度】为"12 mm"，单击【确定】退出设置，如图 1-95 所示。

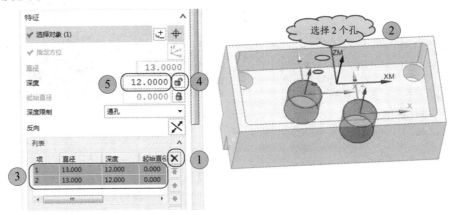

图 1-95　选择 φ13 的两个孔，设置【深度】为"12 mm"

　　(4) 单击　【进给率和速度】图标，设置主轴转速为"500 rpm"，单击转速右侧【计算器】图标，设置进给速度为"100 mmpm"，然后单击【确定】退出对话框，如图 1-96 所示。

　　(5) 单击　【生成】图标，如图 1-97 圆圈 1 所示，计算加工程序，生成【钻孔】加工路径，单击【确定】按钮退出【钻孔】设置，如图 1-97 圆圈 2 所示。

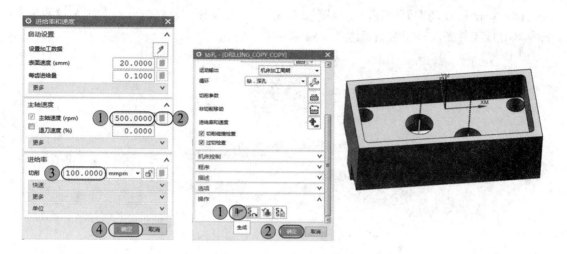

图 1-96　设置转速和进给速度　　　　　　　图 1-97　生成 φ13【钻孔】加工路径

（6）仿真编辑完的加工程序，单击左上方 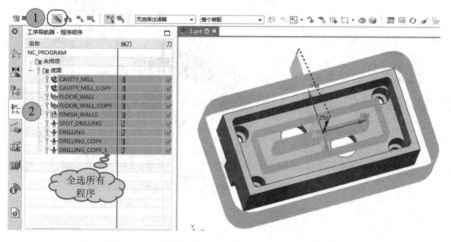【程序顺序】视图，在【程序顺序】视图界面全选所有编完的加工程序，如图 1-98 圆圈 2 所示。

图 1-98　【程序顺序】视图全选所有加工程序

（7）单击【主页】【确认刀轨】图标，如图 1-99 圆圈 1、2 所示，打开【刀轨可视化】对话框，如图 1-100 所示。

图 1-99　单击【确认刀轨】图标

(8) 单击【3D 动态】标签，切换模拟动画为三维立体模型，然后单击下方 ▶【播放】图标，完成工件底面路径的加工仿真，如图 1-100 圆圈 2 所示。

项目一第一序机床仿真

图 1-100　底面加工路径仿真

1.5　开关盒中框的第二序加工

1.5.1　创建父节点组

1. 创建程序组

首先，单击 【程序顺序】视图图标，如图 1-101 圆圈 1 所示，使【工序导航器】切换到【程序顺序】视图。然后，单击屏幕左上角 【创建程序】图标，如图 1-101 圆圈 2 所

项目一第二序上面加工程序编辑

示，弹出【创建程序】对话框，输入名称"上面"，如图 1-101 圆圈 3 所示。单击【确定】如图 1-101 圆圈 4 所示，弹出【程序】对话框，再单击【确定】按键退出【程序】对话框，如图 1-101 圆圈 5 所示。在【程序顺序】视图导航器下增加一个【上面】的程序组，如图 1-101 圆圈 6 所示。

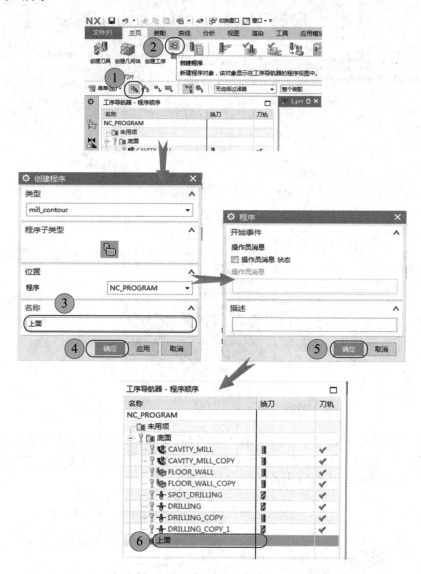

图 1-101　创建【上面】程序组

2. 创建坐标系和几何体

创建第二序加工坐标系和几何体操作步骤如下：

(1) 将【工序导航器】切换到【几何】视图页面，右键单击【MCS_MILL】坐标系，选择【复制】选项，然后右键单击【GEOMETRY】选择【内部粘贴】。这时把第一序的坐标系、几何体和程序全部复制一份，如图 1-102 所示。

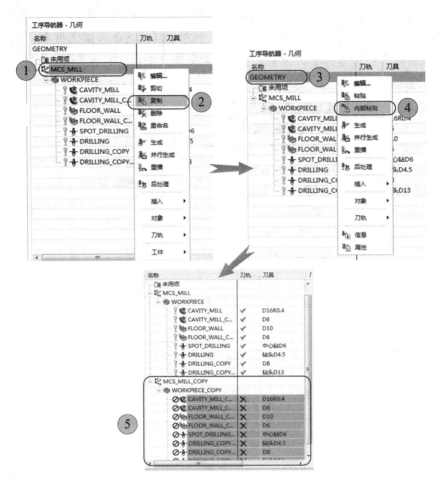

图 1-102 复制第一序的坐标系、几何体和程序

(2) 修改坐标系和几何体名称，右键单击【MCS_MILL】选择【重命名】选项，修改坐标系名称为"底面 MCS"，然后右键单击【WORKPIECE】选择【重命名】选项，修改几何体名称为"底面"，如图 1-103 所示。

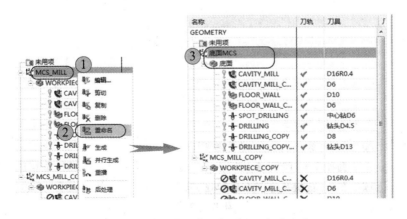

图 1-103 修改第一序坐标系和几何体名称

（3）用同样的方法修改第二序坐标系和几何体名称分别为"上面 MCS"、"上面"。按住键盘【Ctrl】键选中【上面】几何体中所有提示错误的程序，单击右键选择【删除】，删除所有复制过来的程序，如图 1-104 所示。

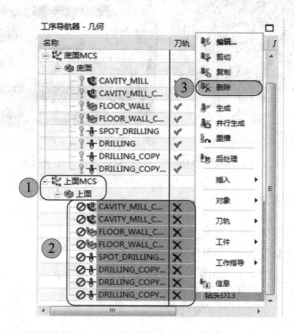

图 1-104　修改第二序坐标系、几何体名称并删除复制的程序

（4）修改坐标系 Z 轴方向，双击 上面MCS【上面 MCS】图标，如图 1-105 圆圈 1 所示。在弹出的【MCS 铣削】对话框中选择 【坐标系对话框】图标，如图 1-105 圆圈 2 所示。

（5）在【坐标系】对话框中选择 动态 图标，如图 1-106 圆圈 3 所示，双击坐标系 ZM 上的箭头使其坐标方向朝上，如图 1-107 圆圈 1 所示。

图 1-105　设置第二序坐标系　　　　　　　图 1-106　选择动态的方式

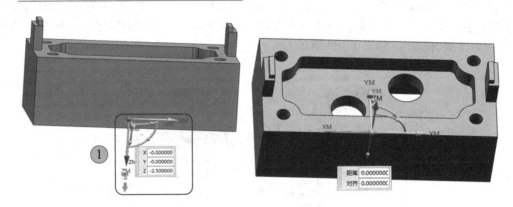

图 1-107 双击 ZM 箭头使其反转创建出二序坐标系

1.5.2 粗加工程序编制

使用 φ16R0.4 钻铣刀粗加工零件上表面程序如下：

(1) 在【程序顺序】视图下，制作第二序粗加工程序。首先，复制【底面】程序组的粗加工程序【CAVITY_MILL】，如图 1-108 圆圈 1 和圆圈 2 所示，右键单击【上面】程序组，选择【内部粘贴】命令，如图 1-108 圆圈 3 和圆圈 4 所示，形成一个新的粗加工程序，如图 1-108 圆圈 5 所示。

(2) 双击打开刚复制过来的 CAVITY_MILL_COPY_1 【型腔铣】程序图标，如图 1-108 圆圈 5 所示，弹出【型腔铣】对话框。在【几何体】选项中把加工几何体选择为新建立的【上面】几何体，这时加工坐标系和几何体就会切换为第二序的坐标系和几何体，如图 1-109 所示。

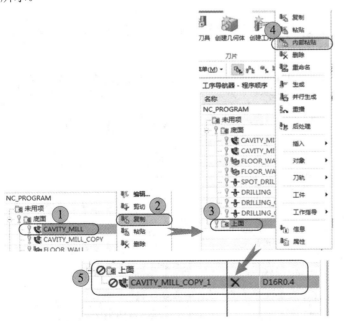

图 1-108 复制【型腔铣】程序到【上面】程序组里　　　　图 1-109 选择第二序几何体

(3) 单击 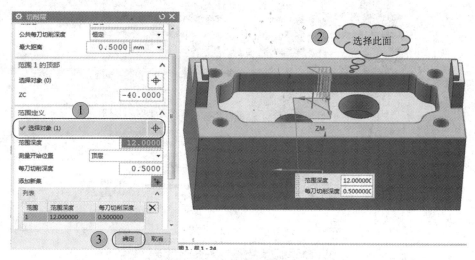【切削层】图标，设置第二序加工深度。由于第一序已经加工过底面大部分地方，因此外框只需要加工到两个凸起台的底面即可。打开【切削层】对话框后直接单击如图 1-110 圆圈 2 所示平面，测得加工【范围深度】为"12 mm"。单击【确定】退出【切削层】对话框。

图 1-110 选择加工底面

(4) 其余参数都不需要修改，直接单击 【生成】图标，如图 1-111 圆圈 1 所示，计算出新的加工路径。单击【确定】退出【型腔铣】，如图 1-111 圆圈 2 所示。

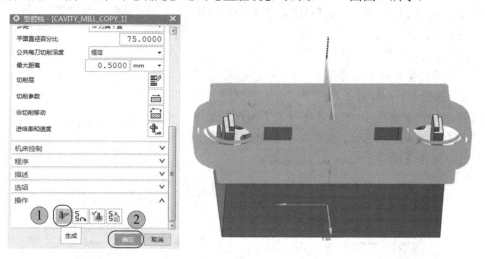

图 1-111 生成第二序粗加工程序

1.5.3 二次开粗程序编制

使用剩余铣的加工方法完成工件的二次开粗程序编制。

(1) 使用 φ10 铣刀粗加工中间凹槽部分。首先复制刚创建的粗加工程序【型腔铣】到【上面】程序组下，形成一个新的【型腔铣】程序，如图 1-112 所示。

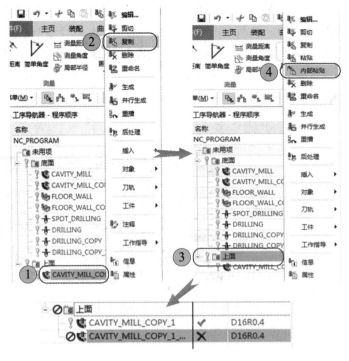

图 1-112　复制新的型腔铣程序

(2) 双击打开新复制的程序，单击【工具】【刀具】下拉箭头，将【刀具】修改为【D10】合金刀，如图 1-113 圆圈 1 所示。

(3) 单击【切削层】图标，设置加工起始和终止深度，首先，单击【范围 1 的顶部】下，⊕【选择对象】图标，选择如图 1-114 圆圈 2 和圆圈 4 所示平面。然后，单击【范围定义】选项中 ⊕【选择对象】图标，选择如图 1-114 圆圈 4 所示平面，测得加工【范围深度】为"7 mm"。单击【确定】按钮，退出设置。

图 1-113　修改刀具为 D10 铣刀

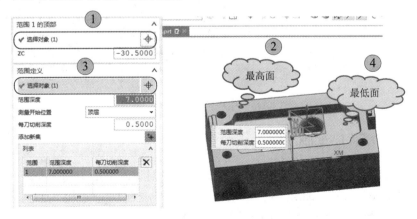

图 1-114　选择加工范围的最高面和最低面

(4) 单击 ▨【切削参数】图标，打开【切削参数】对话框。在【切削参数】对话框中单击【空间范围】选项，【毛坯】中【修剪方式】设置为【轮廓线】，这样生成的程序能自动修剪掉工件范围以外的程序，如图 1-115 所示。

图 1-115　设置【修剪方式】为【轮廓线】

(5) 单击 ▦【生成】图标计算加工程序，生成【型腔铣】加工路径，单击【确定】按钮，退出【型腔铣】设置，如图 1-116 所示。

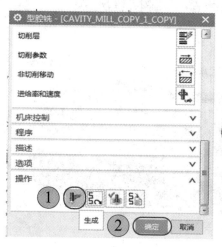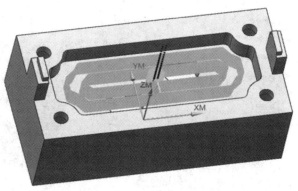

图 1-116　生成 φ10 刀开粗路径

(6) 用和步骤(5)相同的方法编制使用 φ4 合金刀二次开粗内部圆角的程序。首先，建立一把直径 4 mm 的平底刀，刀号为 8 号。复制步骤(5)的【型腔铣】程序，将刀具修改为"D4 平底刀"，设置 Z 向分层【最大距离】为"0.2 mm"每刀，如图 1-117 所示。

(7) 单击【切削参数】图标，在【切削参数】对话框中选择【空间范围】标签，设置参考刀具为【D10】铣刀，如图 1-118 圆圈 3 所示。

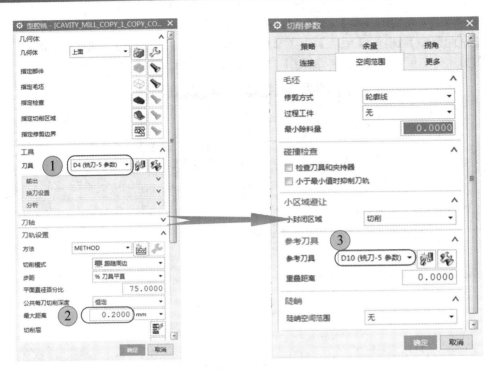

图 1-117　设置【刀具】为【D4(铣刀-5 参数)】　图 1-118　设置【参考刀具】为【D10(铣刀-5 参数)】

(8) 设置【主轴转速】为 "5000 rpm"，单击 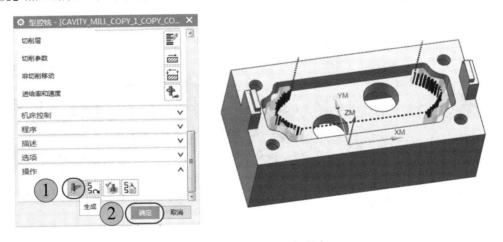【生成】图标计算加工程序，生成【型腔铣】加工路径，单击【确定】按钮退出【型腔铣】设置，如图 1-119 所示。

图 1-119　D4 平底刀开粗程序

1.5.4　精加工底面程序编制

使用 φ10 合金刀完成工件底平面的精加工程序编制。

(1) 在【工序导航器—程序顺序】视图下，制作第二序精加工底面程序。首先，单击右键复制【底面】程序组的精加工底面程序【FLOOR_WALL】，如图 1-120 圆圈 1 和圆圈 2 所示，然后右键单击【内部粘贴】到【上面】程序组下，如图 1-120 圆圈 3 和圆圈 4 所示，形成一个新的【底壁铣】程序，如图 1-120 圆圈 5 所示。

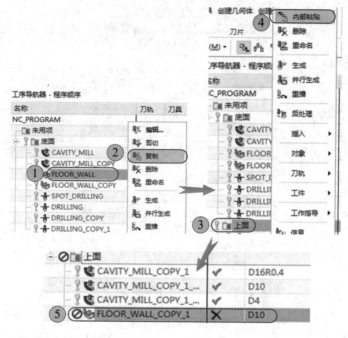

图 1-120 复制第一序的精加工底面程序到第二序程序组中

(2) 双击打开新复制过来的【底壁铣】程序，如图 1-120 圆圈 5 所示，修改参数方法如下：单击【指定切削区域底面】图标，在弹出的【切削区域】对话框中，单击图标删除上次选择的加工底面，如图 1-121 圆圈 1 所示，重新选择新的加工底面共 6 个，如图 1-121 圆圈 3 所示。

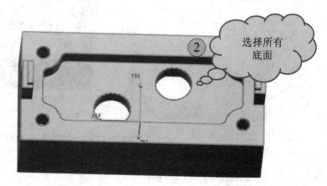

图 1-121 重新选择加工底面

(3) 单击【切削参数】图标，设置【空间范围】标签下的【刀具延展量】为"100%"，如果不设置的话，则生成程序时会出现警报。设置成 100% 后，刀具加工范围会扩大到所有的工件表面，从而避免产生有空面加工不出来的现象，操作步骤如图 1-122 图所示。

(4) 其余参数全部默认原程序不用修改，最后单击【生成】图标计算加工程序，生成【底壁铣】加工路径，单击【确定】按钮退出【底壁铣】设置，如图 1-123 所示。

图 1-122　设置【刀具延展量】为"100%"

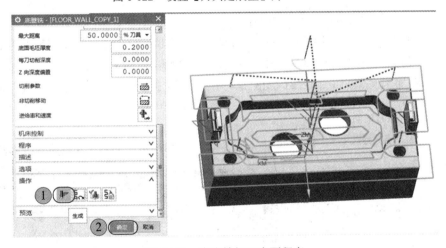

图 1-123　生成精加工底面程序

1.5.5　精加工侧面程序编制

精加工第二序所有侧表面的加工程序方法如下：

(1) 单击【程序顺序】视图制作第二序精加工侧面程序。首先，复制【底面】程序组的精加工侧面程序【FLOOR_WALL_COPY】到【上面】程序组下，形成一个新的【底壁铣】程序，如图 1-124 所示。

(2) 双击打开新复制过来的【底壁铣】程序，如图 1-124 圆圈 6 所示程序，修改参数方法如下：单击打开 ⬛【指定切削区域底面】图标，单击 ✖ 删除原先选择的加工底面，然后重新选择加工底面，共 6 个，参考图 1-121。

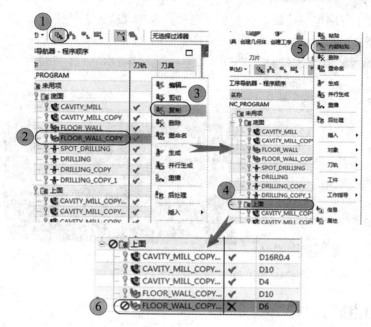

图 1-124 复制第一序的精加工侧面程序到第二序程序组中

（3）单击 📐【切削参数】图标，设置【空间范围】标签里的【刀具延展量】为"100%"。如果不设置的话，则生成程序时会出现警报。设置成 100% 后，刀具加工范围会扩大到所有的工件表面，不会产生有空面没加工的现象，设置如图 1-122 所示。

（4）设置进刀方式为直线进刀。单击 📐【非切削移动】图标，设置【进刀】标签下的【进刀类型】为【线性】，【长度】为刀具直径的"50%"，提刀【高度】为"1 mm"。单击【确定】退出设置，设置参数如图 1-125 所示。

图 1-125 设置进刀方式为直线进刀

（5）其余参数全部使用默认值不用修改，最后单击 🔧【生成】图标，如图 1-126 圆圈

1 所示，计算加工程序生成【底壁铣】加工路径，单击【确定】按钮退出【底壁铣】设置，如图 1-126 圆圈 2 所示。

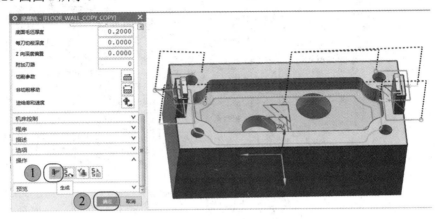

图 1-126　生成精加工侧面程序

1.5.6　精加工侧面清角程序编制

使用 φ4 铣刀编辑精加工零件侧面的清角程序。

(1) 在【工序导航器—程序顺序】视图下制作第二序精加工侧面清角程序。首先，右键复制上一步刚创建的精加工侧面程序【FLOOR_WALL_COPY_COPY】，如图 1-127 圆圈 2 和圆圈 3 所示，然后右键选择【内部粘贴】到【上面】程序组下，如图 1-127 圆圈 4 和圆圈 5 所示，形成一个新的【底壁铣】程序，如图 1-127 圆圈 6 所示。

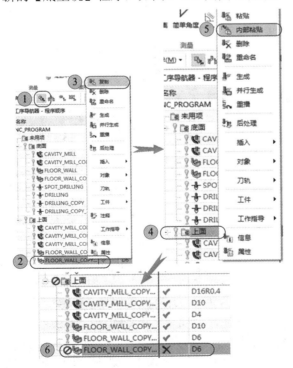

图 1-127　复制上一步的精加工侧面程序到【上面】程序组中

(2) 双击打开新复制过来的【底壁铣】程序。修改参数方法如下：单击打开  【指定切削区域底面】图标，在弹出的【切削区域】对话框中，单击 ✖ 图标删除上次程序的加工底面，重新选择新的加工底面，如图 1-128 圆圈 2 所示。

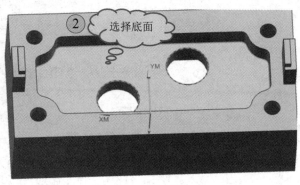

图 1-128 选择型腔底面为加工面

(3) 使用 φ4 刀精加工型腔圆角部分，打开【工具】选项下【刀具】右侧的下拉箭头。展开工具列表，在刀具列表中选择【D4 铣刀】，如图 1-129 圆圈 1 所示。

图 1-129 选择【D4 铣刀】

(4) φ4 铣刀的下刀深度不能很深，否则会产生折刀问题。所以将【刀轨设置】中的【每刀切削深度】设置为"1 mm"，分层加工清角部分，如图 1-130 所示。

图 1-130 设置【每刀切削深度】为"1 mm"

(5) 单击  【切削参数】图标，在【切削参数】对话框中，将【空间范围】标签下的【毛坯】选项设置为【3D IPW】(3D IPW 就是将前序加工后剩余的部分作为本次加工的毛坯使用)。单击【确定】退出设置，操作步骤如图 1-131 所示。

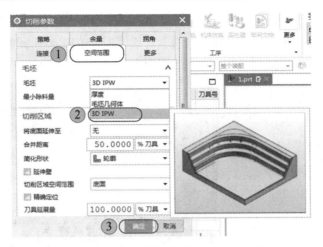

图 1-131 设置工件【毛坯】为【3D IPW】

（6）其余参数全部默认原程序不用修改，最后单击 ⊩【生成】图标计算加工程序，生成精加工清角程序，单击【确定】按钮退出【底壁铣】设置，如图 1-132 所示。

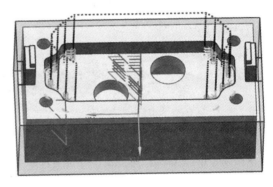

图 1-132 生成精加工清角程序

（7）仿真编辑完所有加工程序，单击左上方 ⓐ【程序顺序】视图图标，如图 1-133 圆圈 1 所示，在【程序顺序】视图界面全选所有编完的加工程序，如图 1-133 圆圈 2 所示。

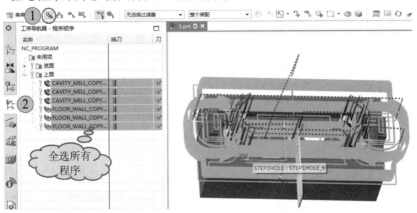

图 1-133 【程序顺序】视图全选所有加工程序

（8）单击【主页】图标，如图 1-134 所示，打开【刀轨可视化】对话框。

图 1-134　单击【确认刀轨】图标

（9）在弹出的【刀轨可视化】对话框中，单击【3D 动态】标签，切换模拟动画为三维立体模型，然后单击下方 ▶【播放】图标，完成工件底面路径的加工仿真，如图 1-135 所示。

项目一第二序机床仿真

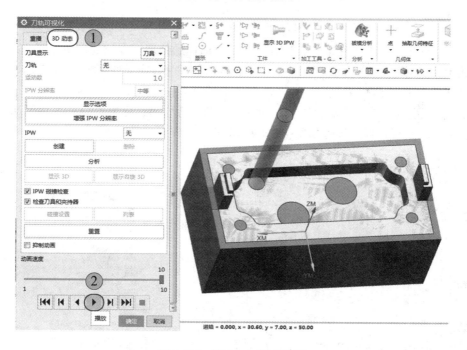

图 1-135　底面加工路径仿真

1.6　生成 G 代码文件

将编辑完的所有加工程序生成 G 代码文件。

(1) 安装 NC 后处理器，单击【菜单】下拉菜单，选择【工具】工具条，向下滚动鼠标滚轮选择【安装 NC 后处理器】选项，操作步骤如图 1-136 所示。

图 1-136　安装 NC 后处理器

(2) 在弹出的【选择后处理器】对话框中，选择后处理文件位置。单击【查找范围】下拉菜单，双击打开光盘目录下【FANUC 0i 后处理】文件夹，在文件夹中选择【FANUC0i.pui】后处理文件，单击【OK】退出设置，操作步骤如图 1-137 所示。

(3) 在弹出的【安装后处理器】对话框中，输入后处理名称为"FANUC0i"，单击【确定】按钮退出设置，操作步骤如图 1-138 所示。

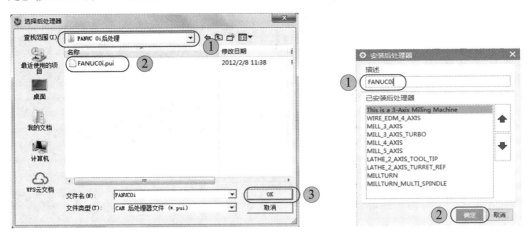

图 1-137　选择后处理文件位置　　　　图 1-138　设置后处理名称为"FANUC0i"

(4) 按住电脑键盘【Ctrl】键，单击选择【底面】程序组下所有程序。单击 【后处理】图标，如图 1-139 所示，弹出【后处理】对话框。

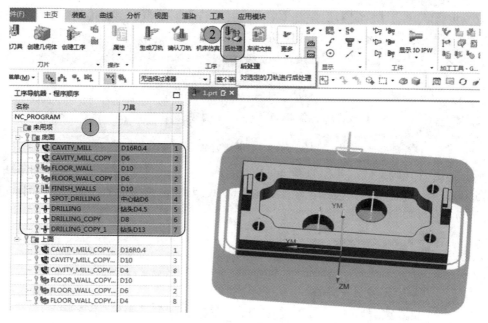

图 1-139　选择底面所有程序单击【后处理】图标

(5) 在【后处理器】选项中选择【FANUC0i】后处理文件，如图 1-140 圆圈 1 所示。单击【输出文件】选项下的 ![icon] 【浏览以查找输出文件】图标，如图 1-140 圆圈 2 所示，弹出【指定 NC 输出】对话框，(首先在 D 盘创建【nc】文件夹，选择 D:\nc 目录，输入文件名为 "1"，如图 1-140 圆圈 3 和圆圈 4 所示，单击【OK】返回【后处理】对话框，如图 1-140 圆圈 5 所示)。确定文件名位置为 D:\nc\1，文件扩展名为 nc，如图 1-140 圆圈 6 所示。单击【确定】退出设置，如图 1-140 圆圈 7 所示。

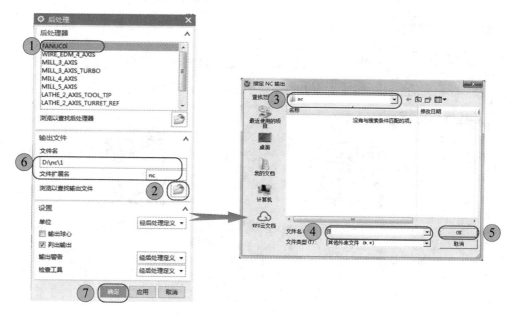

图 1-140　设置后处理文件位置及名称

(6) 步骤(5)单击【确定】后弹出【多重选择警告】对话框，单击【确定】将所有程序输出到一个程序组下显示，如图 1-141 所示。

图 1-141 弹出【多重选择警告】对话框

(7) 弹出 G 代码文件，并在 D 盘 nc 文件夹下生成 1.nc 文件。【底面】程序组下所有程序前面都显示出绿色对勾，表示已经生成 G 代码文件(注：未生成 G 代码文件的加工程序前显示为黄色感叹号)，如图 1-142 所示。

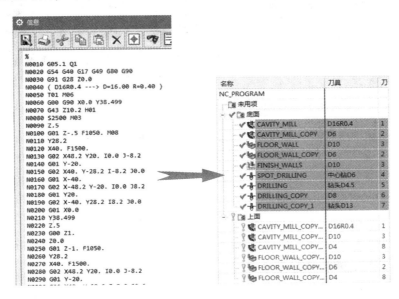

图 1-142 生成 G 代码文件

(8) 以相同的方法生成【上面】程序组中的加工程序。

编程操作视频

项目一第一序编程操作视频　　　　　　项目一第二序编程操作视频

课 后 练 习

按照本项目所学的知识完成课后练习文件的程序编制，课后练习见图 1-143 和光盘"练习"文件夹中 1.prt 文件。

图 1-143　　项目一课后练习图

项目二　批量加工零件的工艺设计与程序编制

案例说明 ✍

　　本项目以光学医疗器械元件加工编程为案例，讲解批量加工制品的装夹工艺方案、多角度加工面零件的加工基准选择、批量加工零件提高切削效率的参数设置、切削刀具选择以及批量加工零件编程的注意事项。

学习目标 ✍

　　通过学习光学医疗器械元件的加工编程，读者应了解和掌握大批量多工序零件的快速装夹方法和使用 NX 软件编辑加工程序的方法。做到举一反三，触类旁通。

学习任务 ✍

　　多工序平面轮廓类零件是机械加工中较常见的加工制品，加工种类较多，且大多批量加工，常应用于电子、医疗、机械、航空等工业领域中。

　　本项目以光学医疗器械元件的加工编程为案例，讲解 NX 软件三轴平面编程。在这个过程中，重点要学会并理解 NX 软件各种平面轮廓程序的加工方法和应用，可以按照不同的方法和工艺生成加工程序。

2.1　批量加工零件的加工工艺规程

　　加工工艺规程是描述每步加工过程的，一般包括被加工的区域、加工类型(平面铣、曲面铣、孔加工等)、工序内容描述、零件装夹、所需刀具及完成加工所必需的其他信息。

　　光电元件外形尺寸为 50 mm×48 mm×31 mm，材料为铝合金 6061，用于此产品批量加工。为便于批量加工并且降低原材料采购成本，购买铝制品原材料时，一般选用外形尺寸稍大一点的毛料自行加工外形尺寸，不需要定制尺寸正好的六面精加工毛坯材料。由于此工件外形尺寸接近 50 mm×50 mm 的正方形，因此在选择原材料时可以选择截面尺寸为 55 mm×55 mm 的铝型材，可自行使用锯床截成每块长度为 37 mm。因此此次选用双边余量大约为 5 mm 左右的毛坯原料，尺寸为 55 mm×55 mm×37 mm 的矩形铝合金型材。

2.1.1　案例工艺分析

　　如图 2-1 和图 2-2 所示为光电元件平面及三维图纸，此工件需要从上、下面和侧孔三方向加工，需要装夹三次零件完成工件的制作。为保证加工后零件的尺寸精度和形位公差，在编程前首先要确定好工件的加工工艺方案。

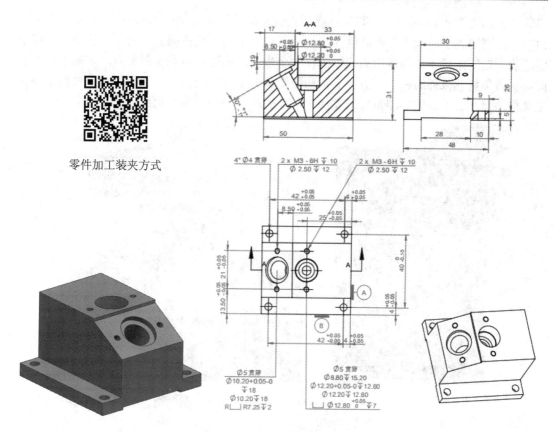

零件加工装夹方式

图 2-1　光电元件平面图纸

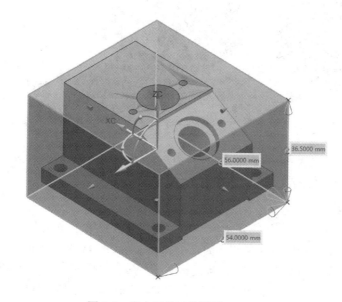

图 2-2　光电元件三维图纸

项目二第一序装夹方式

工序一：

先在虎钳的左侧定钳口上安装一个定位基准块，选择合适的垫铁放在虎钳上，留出钳口的装夹高度为 4.5 mm 左右。将毛坯靠近定位基准块压实并使用虎钳夹紧,确定工件露出钳口部分高度大于被加工零件高度 31 mm,否则可能会出现铣削到虎钳钳口的事故,安装毛坯后一定要测量工件的露出高度以确保安全生产。毛坯安装到虎钳上后,先加工零件上面除斜面和斜孔以外的所有部分。底部和侧孔部分留到后面工序去加工。加工时要注意加工部位的尺寸公差,按照图纸要求完成工件加工,工序一安装示意图如图 2-3 所示。

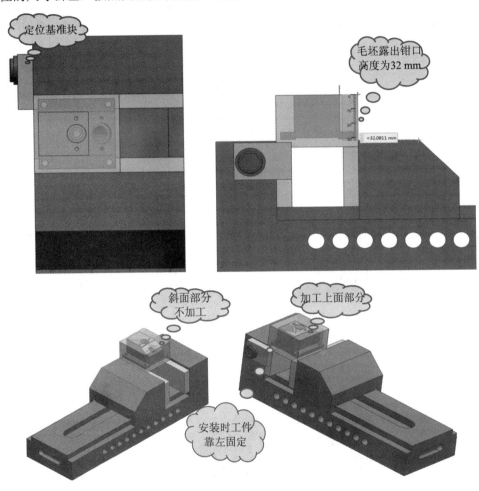

图 2-3   工序一安装示意图

(工序一装配图位置:光盘\例题\2 光电 1 序.prt)

工序二:

先将虎钳上的垫铁撤掉,保留左侧定钳口上安装的定位基准。将第一序加工后的工件底面朝上装夹,使用虎钳夹住工件 30 mm 宽的平面,用第一工序加工的两个 9 mm 宽的平台垫在钳口上表面,将工件靠近左侧定钳口上的固定基准,然后夹紧虎钳完成工件安装。注意在安装前,一定要使用百分表测量两个钳口上表面高度,并保证必须是一致的,不能出现有高度差的问题。如果钳口上表面出现不平行且有高度差的问题,那么加工后的工件底面高度就会出现超差问题。工序二装夹方式示意图如图 2-4 所示。

项目二第二序装夹方式

图 2-4 工序二装夹方式示意图

(工序二装配图位置：光盘\例题\2 光电 2 序.prt)

工序三：

第三序斜面和斜孔加工时使用普通虎钳装夹，并使用三轴加工中心，在不旋转角度的情况下是不可能完成的。为了保证装夹位置的准确和快速装夹，我们为第三序装夹设计了一套专用夹具，以提高装夹速度和定位的准确性。

为便于安装工件并且减少制作胎具时的原材料使用，在此次胎具设计中采用改造虎钳的定动钳口板的方式制作胎具。这样不仅可以使用虎钳现有的夹紧方式固定工件，还能减少制造胎具的制造成本。制作胎具时选用两块尺寸均为 100 mm × 35 mm × 45 mm 的 45#钢作为原材料。按照虎钳上原有的固定螺丝位置，制作螺丝过孔和沉头孔固定到虎钳座上，如图 2-5 所示。

图 2-5 三序胎具固定到虎钳动定钳口上

设计定钳口胎具时要考虑到做出的胎具是否能够把工件轻松地放到钳口里，而且能保证定位面的准确，所设计的钳口应该能够满足加工要求，不会出现负角是否与工件干涉放不进去的问题。从图 2-6 中可以看出，为了便于钳口的加工，可把钳口左右两侧加宽并设计出圆角以减少加工的难度，并在工件的尖角处设计出避让的位置。为避免尖角处配合干涉，此面只保留两个直角边的配合。

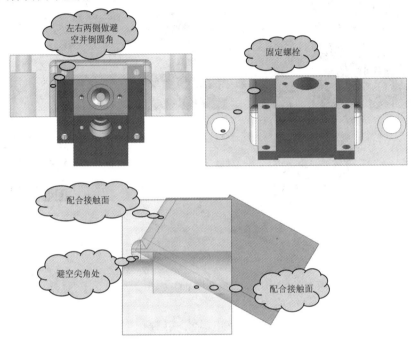

图 2-6　定钳口胎具的设计原理

设计动钳口胎具时也要考虑到做出的胎具是否能够轻松地把工件放到钳口里。从图 2-7 可以看出，为了便于钳口的加工，可把钳口左右两侧加宽并设计出圆角以减少加工的难度，并在工件的尖角处设计出避让的位置。为避免工件和胎具尖角处配合干涉，此面只保留两个侧面和一个斜面的配合。

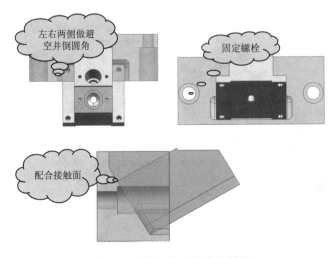

图 2-7　动钳口胎具的设计原理

　　将两钳口固定到虎钳上，然后把工件放在胎具中间，向下压住工件夹紧虎钳。通过虎钳夹紧力夹住工件的两个斜面。工件两个侧面配合时要做到胎具尺寸比工件尺寸双边大 0.04 mm 左右的间隙配合。通过五个面的接触配合来控制工件在胎具中的位置，从而完成工件的三序装夹。工序三工件装夹定位图如图 2-8 所示。

项目二第三序装夹方式

图 2-8　工序三工件装夹定位图

(工序三装配图位置：光盘\例题\2 光电 3 序.prt)

2.1.2　案例加工刀具的选择

　　本例是医疗器械光电元件的数控加工，材料为铝合金 6061，外形尺寸为 50 mm × 48 mm × 31.5 mm。此工件外形不存在凹圆角的结构，因此选择刀具时可以不用考虑凹圆角的问题。但是顶面和侧面两个孔的外型尺寸精度较高，需要精铣加工内圆型腔表面，最小需要精加工的内圆尺寸为 8.8 mm，因此精加工内圆时需要选择 φ8 的铝用合金刀精加工。此工件型腔内外轮廓均由平面构成，首先对其进行粗加工，粗加工时要去除大量的材料。由于加工时刀具承受的力会很大，所以要选取直径比较大的镶片钻铣刀 D16R0.4，这样加工时就不容易发生振刀、断刀、崩刃的现象。按照工件上各孔直径的大小选择相应的中心钻和钻头加工零件上的孔。由于内阶梯孔最小直径为 8.8 mm，所以选用 φ8 合金刀粗加工内孔残余部分。最后使用 φ8 铝用硬质合金刀精铣零件的轮廓、内孔和底面。

　　按照零件加工工艺方案和切削刀具的选择方式，合理安排零件的加工工艺过程。按照先粗后精、先面后孔、基准统一的原则设计本案例的加工工艺过程单。如表 2-1 所示。

表 2-1 光电元件工艺过程单

工序号	顺序号	加工机床	工序内容	工序名	刀具名称
1	1	加工中心	粗加工	型腔铣	D16R0.4
1	2	加工中心	二次开粗	剩余铣	D8
1	3	加工中心	底面精加工	底壁铣	D8
1	4	加工中心	精加工上层外形侧壁	精铣壁	D8
1	5	加工中心	精加工下层外形侧壁	精铣壁	D8
1	6	加工中心	精加工上层内孔侧壁	精铣壁	D8
1	7	加工中心	精加工中间内孔侧壁	精铣壁	D8
1	8	加工中心	精加工下层内孔侧壁	精铣壁	D8
1	9	加工中心	精加工左侧边侧壁	精铣壁	D8
1	10	加工中心	打中心钻	定心钻	中心钻 D6
1	11	加工中心	加工φ2.5 的孔	钻孔	钻头 D2.5
1	12	加工中心	加工φ4 的孔	钻孔	钻头 D4
1	13	加工中心	加工φ5 的孔	钻孔	钻头 D5
1	14	加工中心	加工 M3 螺纹孔	攻丝	M3 丝锥
2	15	加工中心	粗加工	型腔铣	D16R0.4
2	16	加工中心	底面精加工	底壁铣	D8
2	17	加工中心	精加工侧壁	底壁铣	D8
3	18	加工中心	粗加工	型腔铣	D8
3	19	加工中心	底面精加工	底壁铣	D8
3	20	加工中心	精加工侧壁	精铣壁	D8
3	21	加工中心	打中心钻	定心钻	中心钻 D6
3	22	加工中心	加工φ2.5 的孔	钻孔	钻头 D2.5
3	23	加工中心	加工φ10.2 的孔	钻孔	钻头 D10.2
3	24	加工中心	加工φ5 的孔	钻孔	钻头 D5
3	25	加工中心	加工 M3 螺纹孔	攻丝	M3 丝锥

2.2 打开模型文件进入加工模块

打开随书配套光盘，在"例题"文件夹中打开模型文件，并且进入加工模块。

(1) 启动 NX 12.0，单击左上角 【打开】按钮，在【打开】对话框中选择光盘"例题"文件夹中 2.prt 文件，如图 2-9 所示。

光学医疗器械元件加工程序编制

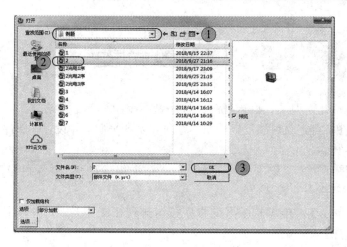

图 2-9　打开光盘例题文件夹中 2.prt 文件

(2) 单击【应用模块】按钮，再单击【加工】按钮(也可直接单击键盘快捷键 Ctrl+Alt+M)启动 NX 12.0 加工模块，如图 2-10 所示。

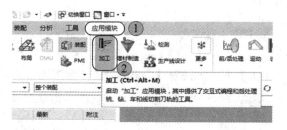

图 2-10　进入 NX12.0 加工模块

(3) 打开【加工环境】，默认选择【CAM 会话配置】中【cam_general】，在【要创建的 CAM 组装】选项中选择【mill_planar】平面铣按钮，单击【确定】按钮进入平面铣加工界面，如图 2-11 所示。

图 2-11　设置【加工环境】对话框

2.3　建立父节点组

父节点组包括几何视图、机床视图、程序顺序视图和加工方法视图。

(1) 几何视图：可定义"加工坐标系"方向和安全平面，并设置"部件""毛坯"和"检查"等几何体参数。

(2) 机床视图：可定义切削刀具，如铣刀、钻头和车刀等，并保存与刀具相关的数据，以用作相应后处理命令的默认值。

(3) 程序顺序视图：能够把编好的程序按组排列在文件夹中，并按照从上到下的先后顺序排列加工程序。

(4) 加工方法视图：用来定义切削方法类型(粗加工、精加工、半精加工)。例如，"内公差""外公差"和"部件余量"等参数在此设置。

2.3.1　创建加工坐标系

在【几何】视图菜单中创建加工坐标系的操作步骤如下：

(1) 将【工序导航器】切换到【几何】视图页面，如图 2-12 所示。

(2) 右击【MCS_MILL】图标，选择【重命名】，设置工序一坐标系名称为【MCS_1】，如图 2-13 圆圈 1 所示，修改方法参考如图 1-103 所示，双击打开【MCS 铣削】对话框。

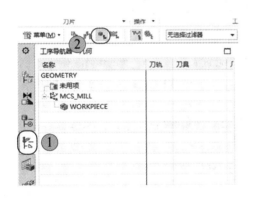

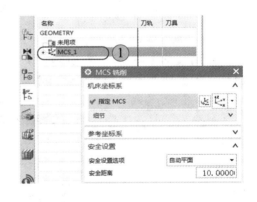

图 2-12　【几何】视图切换　　　　　图 2-13　【MCS 铣削】对话框

(3) 由于毛坯材料没有经过机加工处理，表面形状可能不规则，因此为保证各面加工留量均匀，把坐标系放在毛坯中间位置。单击 【坐标系对话框】图标，如图 2-14 圆圈 1 所示。选择【对象的坐标系】选项，如图 2-14 圆圈 2 所示。设置坐标为所选择平面的中心点位置。单击工件上表面方框，自动捕捉出工件的上表面中心点坐标位置，如图 2-14 圆圈 3 所示。在抬刀安全平面没有干涉物的情况下，可以选择默认状态【自动平面】，安全距离为"10 mm"。如有干涉物，则可把安全距离设置为"50 mm～100 mm"。最后单击【确定】按钮退出【MCS 铣削对话框】，如图 2-14 圆圈 4 和圆圈 5 所示。

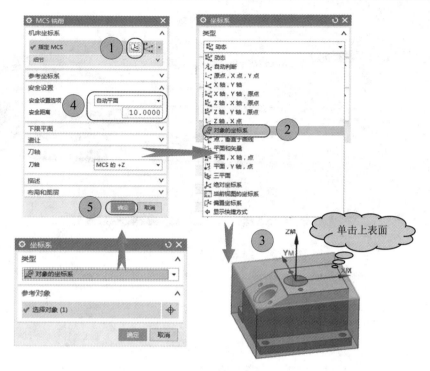

图 2-14　设置工序一加工坐标系

2.3.2　创建部件几何体

在【几何】视图菜单中创建加工部件几何体、零件毛坯、检查几何体的操作步骤如下：

(1) 右击 ⬢ WORKPIECE 图标，选择【重命名】，设置工序一部件名称为"1"，如图 2-15 圆圈 1 所示，双击打开【工件】对话框，如图 2-15 所示。

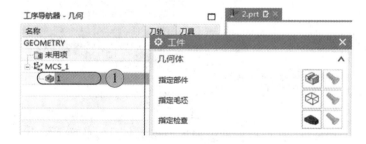

图 2-15　打开【工件】对话框

(2) 单击 ⬢【选择或编辑部件几何体】图标，如图 2-16 圆圈 1 所示，弹出【部件几何体】对话框。单击被加工工件使其成橘黄色，然后单击【确定】按钮退出【部件几何体】对话框。

(3) 单击 ⬡【选择或编辑毛坯几何体】图标，如图 2-16 方框 1 所示，弹出【毛坯几何体】对话框，在【类型】选项中选择【几何体】选项，如图 2-17 方框 2 所示。单击图中给定的毛坯，如图 2-17 方框 3 所示，单击【确定】按钮，如图 2-17 方框 4 所示，返回【工件】对话框，然后再单击【确定】按钮退出【工件】对话框设置，如图 2-17 方框 5 所示。

图 2-16　创建部件几何体

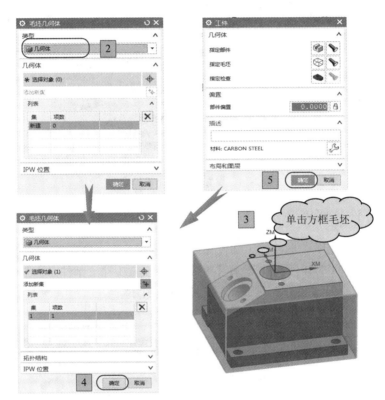

图 2-17　建立毛坯几何体

(4) 在选择毛坯后，毛坯方框在编辑加工程序中就不再使用，这时我们可以把它隐藏起来，避免后序操作中产生误操作现象。其具体方法为：单击毛坯方框，按键盘快捷键 Ctrl + B。如果想使其继续显示，则使用键盘快捷键 Ctrl + Shift + K，然后单击选择要恢复的图形退出【显示】设置。

2.3.3　创建刀具

在【机床】视图下创建加工刀具步骤如下：

(1) 单击 【机床】视图图标，将【工序导航器】切换到【机床】视图页面，如图 2-18 所示。

(2) 单击 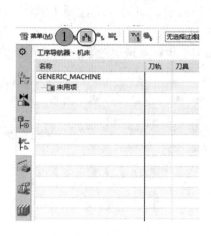 【创建刀具】图标，如图 2-19 圆圈 2 所示，弹出【创建刀具】对话框，如图 2-19 所示。

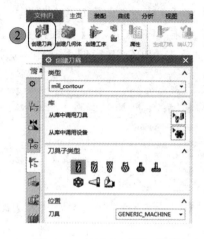

图 2-18　【机床】视图　　　　　　　　　图 2-19　【创建刀具】对话框

(3) 单击 【MILL】图标，创建平底刀。刀具【名称】位置输入 "D16R0.4" (代表直径为 16 mm、圆角半径为 0.4 mm 的镶片钻铣刀)，如图 2-20 所示。

(4) 在【铣刀-5 参数】对话框中，直径设置为 "16"，下半径设置为 "0.4"，【刀具号】、【补偿寄存器】、【刀具补偿寄存器】三项均设置为 "1" (此数值代表刀具、刀具半径补偿和刀具长度补偿号，为避免发生撞机问题，最好设置为相同数字)。单击【确定】按钮完成刀具建立，如图 2-21 所示。

其他刀具在编辑加工程序前，按照给定参数自行设置。

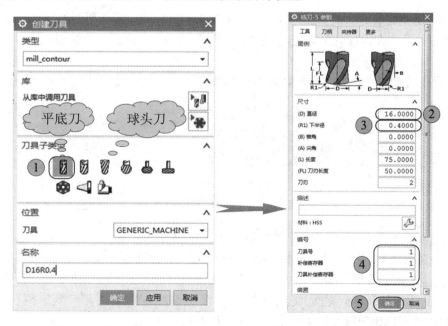

图 2-20　创建刀具　　　　　　　　　图 2-21　【铣刀-5 参数】对话框设置

2.3.4 创建程序组

在【程序顺序】视图中创建加工程序组文件夹，操作步骤如下：

(1) 将【工序导航器】切换到【程序顺序】视图页面，如图 2-22 圆圈 1 所示。

(2) 双击【PROGRAM】程序组文字，如图 2-22 方块 1 所示，修改文件名为"1"(或使用右键单击【PROGRAM】程序组，选择【重命名】也可实现更改名称)，如图 2-23 方块 2 所示。

图 2-22 【工序导航器】切换到【程序顺序】视图页面

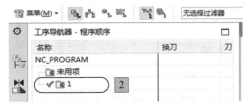

图 2-23 双击修改程序组名称

(3) 保存文件。

2.4 批量加工零件的第一序加工

由前面的加工分析得出此工件的第一序加工为零件正面朝上，用虎钳夹住毛坯多出的部分，加工除斜面和斜孔以外所有的面。以下为第一序加工的程序编制方法。

2.4.1 粗加工程序编制

这里使用型腔铣的加工方法完成零件的粗加工程序编制。为了更好地体现出加工程序的先后顺序，我们在编程时全部使用【程序顺序】视图来完成程序的编制。具体操作步骤如下：

(1) 单击【创建工序】图标，如图 2-24 圆圈 1 所示，弹出【创建工序】对话框。

图 2-24 单击【创建工序】图标

(2) 在【类型】下拉菜单中选择【mill_contour】曲面铣选项，如图 2-25 圆圈 2 所示。选择 【型腔铣】图标，如图 2-26 圆圈 3 所示。在【程序】下拉菜单中选择新建的【底面】程序组，【刀具】下拉菜单中选择【D16R0.4】的镶片钻铣刀，【几何体】下拉菜单中选择建立好的【WORKPIECE】几何体，在【名称】栏中可以按照加工要求输入一个程序名称，本例在这里不做专门修改，按照默认名称填写即可，然后单击【确定】按钮，如图 2-26 圆圈 4 和圆圈 5 所示。

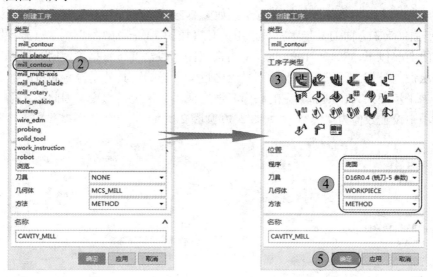

图 2-25　选择【mill_contour】曲面铣选项　　　　图 2-26　创建【型腔铣】工序

(3) 在弹出的【型腔铣】对话框中，在【几何视图】中正确设定【WORKPIECE】。在进入型腔铣时，【指定部件】和【指定毛坯】选项应显示为灰色，当右侧 🔧【显示】图标为彩色时，单击可显示已选择的几何体部件。如果进入型腔铣后还能选择部件和毛坯，则说明【几何】视图中的【WORKPIECE】没有设定，或进入程序前没有选择几何体为【WORKPIECE】。型腔铣编程中一般不用设置【指定切削区域】，【指定检查】和【指定修剪边界】根据零件加工需求设置，在本例中不需要设置，如图 2-27 所示。

(4) 单击【工具】菜单下拉箭头，显示已选择的加工刀具为 D16R0.4 镶片钻铣刀(注：此项工作前序选择正确的情况下可忽略)，如图 2-28 所示。

(5) 单击【刀轴】菜单下拉箭头，显示出默认刀轴为+ZM 轴，三轴加工中心刀轴一般使用+ZM 轴，只有在使用多轴机床加工时才会修改此项(注：此选项三轴加工编程时不用选择，使用默认设置即可)，如图 2-28 所示。

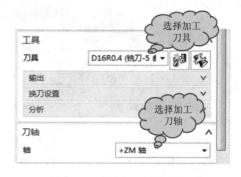

图 2-27　几何体设置对话框　　　　　　　　图 2-28　工具和刀轴选项

(6)【刀轨设置】为型腔铣参数设置的主要内容。切削模式选项中一般常用 🗔 跟随部件和  跟随周边 两种方式，【跟随部件】适合加工开放轮廓的工件，可以使刀具从外向内加工，

并从工件外下刀。【跟随周边】更适合加工封闭轮廓的工件，可以使刀具从内向外加工，减少型腔加工时的下刀位置变化。此案例工序一加工时凸台开放轮廓加工面较多，所以在编程中更适合选择【跟随部件】的加工方法，如图 2-29 所示。

(7) 粗加工 XY 方向刀具步距一般使用刀具直径的 70%～80%，精加工时刀具步距使用刀具直径的 50%以下，80%～100%的 XY 方向步距一般情况下不推荐使用。若刀具步距太大，则每次切削时相当于满刀切削，刀具受力过大会影响刀具使用寿命和机床加工精度；其次，若加工步距太大，则加工完的底面光洁度低，会出现接刀痕。因此在【平面直径百分比】选项中设置数值为"75%"，如图 2-30 圆圈 2 所示。

图 2-29　选择【跟随部件】

图 2-30　XYZ 方向步距量设置

(8) D16R0.4 钻铣刀粗加工时的 Z 方向步距一般情况下取值 0.3 mm～0.7 mm 每层。【最大距离】选项设置的就是刀具 Z 方向的每层步距量，因此设置中间数值为"0.5 mm"每层，如图 2-30 圆圈 3 所示。

(9) 单击 ⚏ 【切削参数】图标，如图 2-30 圆圈 4 所示，打开【切削参数】对话框。

(10) 在【切削参数】对话框的【策略】标签下设置【切削顺序】为【深度优先】，【深度优先】会按照不同区域分别由上往下加工，可以减少抬刀和过刀路径，减少加工时间。【层优先】会按照同一深度在不同区域跳刀加工，从而会增加很多抬刀路径，增加加工时间。一般情况下优先选择【深度优先】，如图 2-31 所示。

图 2-31　【策略】标签设置

(11) 在【切削参数】对话框中选择【余量】标签，设置【部件侧面余量】参数为"0.2 mm"(注：粗加工刀具余量一般设置为"0.2 mm")，如图 2-32 所示。

(12) 在【切削参数】对话框中选择【连接】标签，由于选择的是【跟随部件】的切削方式，因此在【连接】标签下就会增加【开放刀路】的切削方法选择。在【开放刀路】选项下有两种切削方法，分别为【保持切削方向】和【变换切削方向】。【保持切削方向】是指加工刀路会按照事先设定好的顺铣或逆铣的方法铣削，在每一刀切削后会使用快速提刀回到起始位置继续加工，这样的好处是加工程序的每刀切削顺逆铣是相同的，但缺点是抬刀过多会影响加工效率。【变换切削方向】是指加工刀路会按照最短的距离连接刀路，形成往复式的加工路径。这种加工方法的优点是减少了不必要的抬刀，提高了加工效率。其缺点是在加工中会形成以一刀顺铣一

图 2-32 【余量】标签设置

刀逆铣的方式往复加工，这种方法对刀具的寿命和被加工工件表面光洁度都会有一定的影响，但是粗加工不需要考虑加工表面的质量，只需要提高加工效率即可，因此我们常使用的加工方法为【变换切削方向】。设置完成后单击【确定】按钮退出【切削参数】对话框，具体设置如图 2-33 所示。

图 2-33 开放刀路设置

(13) 单击▨【非切削移动】图标，如图 2-34 圆圈 1 所示，打开【非切削移动】对话框。

(14) 设置进刀参数首先要设置【封闭区域】下刀参数，型腔内下刀一般采用螺旋下刀的方式。螺旋下刀加工原理：在型腔尺寸能够满足螺旋下刀时，选用螺旋下刀的方式进刀。在型腔尺寸不能满足螺旋下刀时，选用斜线下刀或直线下刀的方式进行。螺旋下刀【直径】

选用刀具直径的"50%",【斜坡角度】设置为"5°",【高度】设置为"1 mm",【最小斜坡长度】设置为"0"(注:加工刀具能够直线下刀时可以输入数字为 0;如果加工刀具不能直线下刀,此数字最小不能小于 50,否则会发生撞机事故),如图 2-35 圆圈 2 所示。【开放区域】设置进刀长度为刀具直径的"50%",抬刀【高度】设置为"1 mm",以尽量减少抬刀距离,如图 2-35 圆圈 3 所示。

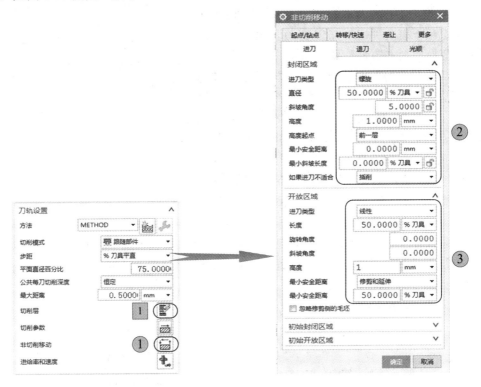

图 2-34 选择非切削移动   图 2-35 进刀参数设置

(15) 单击【非切削移动】对话框【转移/快速】标签设置快速抬刀高度。为提高加工速度减少抬刀高度,把【区域之间】和【区域内】的【转移类型】都改为【前一平面】,并且把抬刀【安全距离】都设置为"1 mm"。这样做的好处是能减少不必要的抬刀,节约加工时间。但大部分快速移刀都是在工件零平面以下进行,所以要求机床的 G00 运动必须是两点间的直线运动,不能是两点间的折线运动,否则会发生撞机事故。加工前一定要在 MDI 下输入"G00 走斜线观察机床"的移动方式,如果不对,则需要修改机床参数,或者按照 NX 12.0【转移/快速】的初始设置方法,把【转移类型】全部设置为【安全距离-刀轴】。最后,单击【确定】按钮退出【非切削移动】对话框,具体参数设置如图 2-36 所示。

图 2-36 【转移/快速】标签设置

(16) 单击 【切削层】图标，如图 2-34 方框 1 所示，进入【切削层】对话框。

(17) 在【切削层】对话框中多次单击 ✕ ，将列表中数值全部删除，如图 2-37 圆圈 1 所示。

图 2-37　删除列表数值

(18) 单击【范围定义】里的【选择对象】，选择零件如图 2-38 圆圈 1 所示位置。测得加工【范围深度】为 "31 mm"，单击【确定】按钮退出【切削层】对话框。

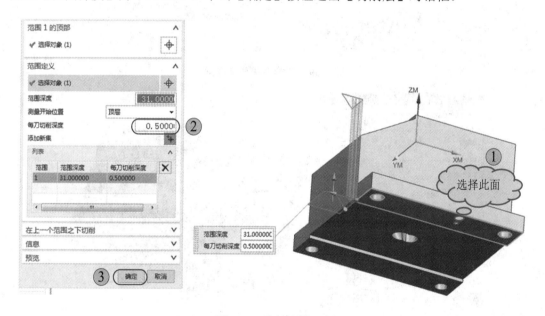

图 2-38　选择切削深度

(19) 单击 【进给率和速度】图标，如图 2-39 圆圈 1 所示，打开【进给率和速度】对话框。设置【主轴速度】为 "2500 rpm"(注：输入 "2500" 后一定要按后面的【计算器】图标，如圆圈 3 所示，否则会报警)。【进给率】的【切削】选项设置为 "1500 mmpm"，【更多】选项中【进刀】为 "70%切削"，然后单击【确认】退出【进给率和速度】对话框，如图 2-39 所示。

(20) 单击 【生成】图标计算加工路径。单击【确定】按钮退出型腔铣设置，如图 2-40 所示。

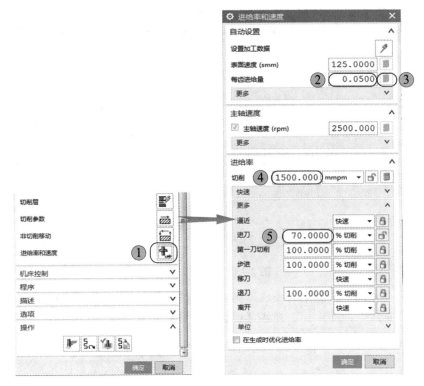

图 2-39　进给率和速度设置

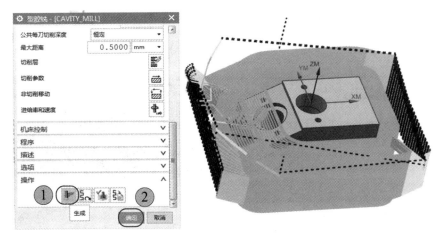

图 2-40　生成刀具路径图

2.4.2　二次开粗程序编制

因为顶面阶梯孔均为平底孔，而且尺寸公差要求较高，所以不能使用钻头直接打孔成型，而只能使用平底铣刀铣削内圆。最小铣削圆孔直径为 8.8 mm，所以选用直径为 8 mm 的平底刀进行圆孔毛坯的二次开粗工作。具体操作步骤如下：

(1) 由【工序导航器】切换至【机床】视图，然后建立一把直径为 8 mm 的硬质合金刀，刀号为 2 号，刀具名称为 D8，参考如图 2-18~图 2-21 所示方法。

(2) 使用剩余铣开粗圆孔型腔，在【机床】视图中右键单击刚编好的【CAVITY_MILL】程序，选择【复制】选项，如图 2-41(a)所示，然后右键单击【D8】刀具，选择【内部粘贴】，如图 2-41(b)所示。形成一个新的提示错误的【型腔铣】程序，如图 2-42 所示。

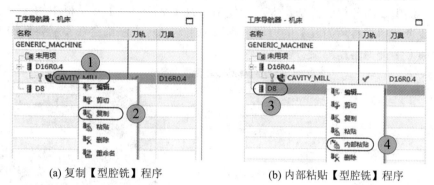

(a) 复制【型腔铣】程序　　　　　　(b) 内部粘贴【型腔铣】程序

图 2-41　【型腔铣】程序设置

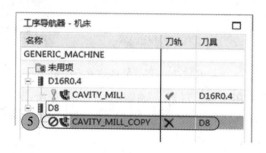

图 2-42　新的【型腔铣】程序

(3) 双击打开新复制过来的【型腔铣】程序。设置 Z 向分层【最大距离】为"0.2 mm"，如图 2-43 所示。

图 2-43　设置 Z 向分层为"0.2 mm"每刀

(4) 单击 ⊞【指定修建边界】图标，如图 2-44 圆圈 1 所示，弹出【修建边界】对话框。在工件表面选择圆型腔外边界线作为修剪边界。【修剪侧】选择为【外侧】，就是去除边界线以外的刀具路径，只保留边界线以内的刀具路径。【平面】选择【自动】，即让软件根据所选的边界线位置自己设定起始平面。具体操作步骤如图 2-44 所示。

在此案例中二次开粗只需要加工顶部圆孔部分，其余部分都不需要二次开粗加工，所以在这个案例中二次开粗的方法不太适合使用修改【空间范围】对话框中的【使用基于层的】选项或者使用【参考刀具】的方法，因为使用这两种方法都会在不需要精加工的斜面部分产生很多多余刀路。

图 2-44　设置修剪边界方法

(5) 单击 ![]【进给率和速度】图标，如图 2-45 方块 1 所示，弹出如图 2-46 所示对话框，设置【主轴速度】为 "4000 rpm"，单击转速右侧 ![] 【计算器】图标，最后单击【确定】按钮退出【进给率和速度】对话框。

(6) 单击 ![]【生成】图标计算加工程序，生成【型腔铣】加工路径，单击【确定】按钮退出【型腔铣】设置。

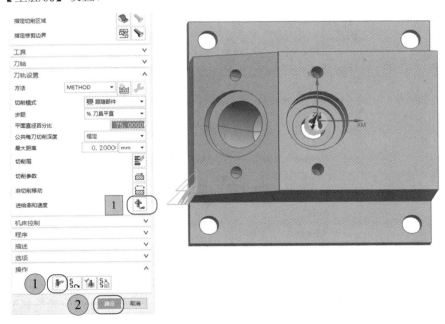

图 2-45　生成二次开粗刀具路径

图 2-46 设置进给率和速度

2.4.3 精加工底面程序编制

使用 φ8 平底刀编制加工底面程序操作如下:

(1) 单击 【创建工序】图标生成创建精加工底面程序。在【类型】下拉菜单中选择【mill_planar】平面铣选项,【工序子类型】选项中选择 【底壁铣】加工方法,【程序】选项中选择【1】程序组,【刀具】选项中选择上一步已经创建的【D8】铣刀,【几何体】选项中选择【1】,最后单击【确定】按钮进入【底壁铣】对话框,操作步骤如图 2-47 圆圈标记所示。

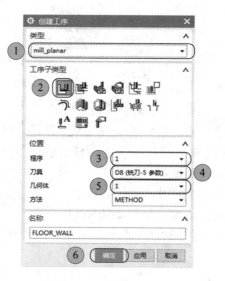

图 2-47 创建精加工底面工序

(2) 单击 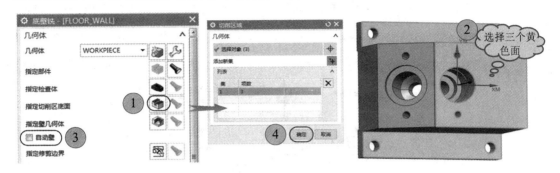 【指定切削区域底面】图标选择要加工的底面。单击型腔底面，如图 2-48 圆圈 2 所示，单击【确定】按钮退出【切削区域】对话框，操作步骤如图 2-48 所示。

图 2-48　【切削区域】对话框选择底面

(3) 勾选 🔲 自动壁　【自动壁】图标前的方框，使其自动捕捉和已选择底面相邻的工件侧壁为壁几何体，如图 2-48 圆圈 3 所示。

(4) 因为被加工表面为开放平面，更适合使用【跟随部件】的加工方法，所以设置【切削模式】为【跟随部件】。精加工时 XY 方向的步距量应小于刀具直径的 50%。在软件中【最大距离】默认为"50%"，所以不需要修改。【底面毛坯厚度】设置为粗加工时底面的余量，在此选项框中输入"0.2 mm"，操作步骤如图 2-49 所示。

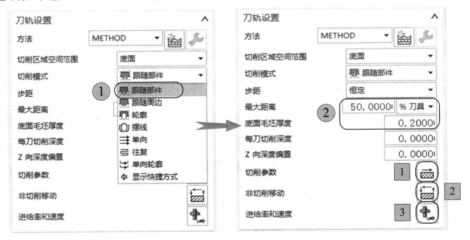

图 2-49　【刀轨设置】对话框选择

(5) 单击 📇 【切削参数】图标，如图 2-49 方块 1 所示。打开【切削参数】对话框，单击上方【余量】标签，设置【壁余量】为"0.1 mm"(注意：由于此程序是精加工底面程序，所以【最终底面余量】应为"0 mm"。工件壁的精加工应由专门的程序去执行，所以给精加工壁的程序留有单边 0.1 mm 的加工余量)。减小公差的数值提高加工精度，设置【内公差】和【外公差】均为"0.01 mm"，如图 2-50 所示。

(6) 在【连接】标签中设置开放刀路的加工方法为【变换切削方向】，然后单击【确定】按钮退出【切削参数】对话框。

(7) 单击【非切削移动】图标，如图 2-49 方框二所示，弹出【非切削移动】对话框，单击左上方【进刀】标签，在【进刀】标签中设置进刀参数，如图 2-51 所示。

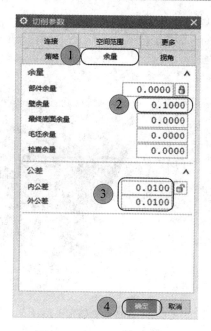

图 2-50 加工【余量】设置

(8) 单击 【进给率和速度】图标，如图 2-49 方框 3 所示，弹出【进给率和速度】对话框，设置【主轴速度】为"3500 rpm"，单击右侧 【计算器】图标。设置【进给率】中的【切削】参数为"1000 mmpm"。设置【更多】选项中【进刀】为"70%"降低进刀速度，然后单击【确定】按钮退出【进给率和速度】对话框，如图 2-52 所示。

图 2-51 进刀参数设置

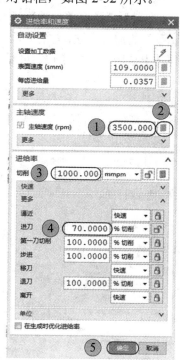

图 2-52 【进给率和速度】对话框

（9）单击 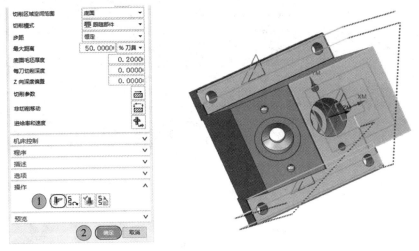【生成】图标计算加工程序，生成【底壁铣】加工路径，单击【确定】按钮退出【底壁铣】设置，如图 2-53 所示。

图 2-53　生成底壁铣加工路径

2.4.4　精加工侧面程序编制

使用 ϕ8 的合金刀精加工零件内外轮廓的编制方法如下：

（1）使用 ϕ8 合金刀精加工外形轮廓。单击 【创建工序】图标，弹出【创建工序】对话框，在【类型】下拉菜单中选择【mill_planar】平面铣选项，【工序子类型】选项中选择 【精铣壁】的加工方法，【程序】选项中选择【1】程序组，【刀具】选项中选择【D8】铣刀，【几何体】选项中选择【1】，最后单击【确定】按钮进入【精铣壁】对话框，操作步骤如图 2-54 所示。

图 2-54　创建【精铣壁】程序

(2) 单击 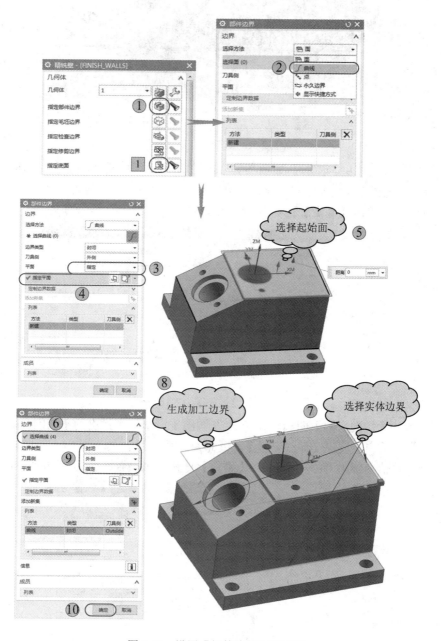【指定部件边界】图标，如图 2-55 圆圈 1 所示，打开【部件边界】对话框。在边界中的【选择方法】选项中选择【曲线】的方式创建加工边界，设置边界类型为【封闭】，刀具侧为【外侧】，平面选择【指定】，如图 2-55 圆圈 9 所示。单击工件上表面为加工起始平面，然后手动选择轮廓外圈线在指定平面生成新的加工边界，单击【确定】退出设置。

图 2-55　设置【部件边界】对话框

(3) 单击 【指定底面】图标，如图 2-55 方块 1 所示，弹出【平面】对话框。设置加工底面，选择底部平台上表面为加工底面，如图 2-56 所示。

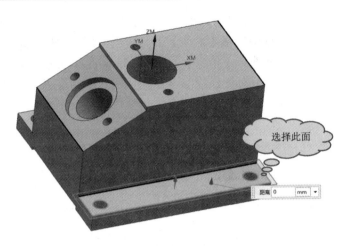

图 2-56   选择加工底面

(4) 单击 ≣ 【切削层】图标，如图 2-57 圆圈 1 所示，弹出【切削层】对话框，在【类型】下拉菜单中选择【恒定】的切削方式，每层的高度是恒定的，【每刀切削深度】【公共】设置为"13 mm"，单击【确定】按钮退出设置，如图 2-57 所示。

图 2-57   设置分层高度为"13 mm"

(5) 单击 ▨ 【切削参数】图标，如图 2-59 圆圈 1 所示，在弹出的【切削参数】对话框中的【余量】选项中，设置【部件余量】和【最终底面余量】全部为"0"，设置【内公差】和【外公差】数值全部为"0.01"，单击【确定】按钮退出设置，如图 2-58 所示。

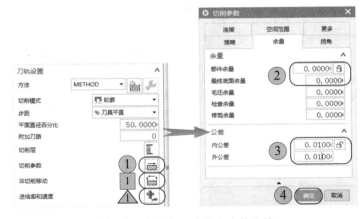

图 2-58   设置加工余量和内外公差

(6) 单击 【非切削移动】图标，如图 2-58 方框 1 所示，在【进刀】标签下设置【进刀类型】为【圆弧】进刀并且打开半径补偿功能，具体设置方法如图 2-59 所示。

图 2-59 设置【非切削移动】对话框

(7) 单击 【进给率和速度】图标，如图 2-58 三角 1 所示，设置【主轴速度】为"3500 rpm"，【进给率】的【切削】参数为"500 mmpm"(精加工侧壁时走刀速度不宜过快，应控制在"500 mmpm"以下，否则加工表面粗糙度过大影响加工质量)，然后单击【确定】按钮退出【进给率和速度】对话框，如图 2-60 所示。

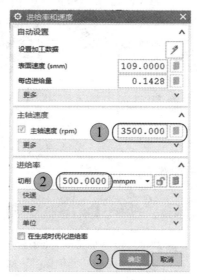

图 2-60 设置主轴转速和进给速度

(8) 单击 【生成】图标，计算加工程序，生成【精铣壁】【精加工上台侧壁】路径，单击【确定】按钮退出【精铣壁】设置，如图 2-61 所示。

(9) 在【机床】视图中右键单击刚编好的【FINISH_WALLS】程序，选择【复制】选项，然后右键单击【D8】刀具选择【内部粘贴】，形成一个新的提示错误的【精铣壁】程序，具体操作方法如图 2-62 所示。

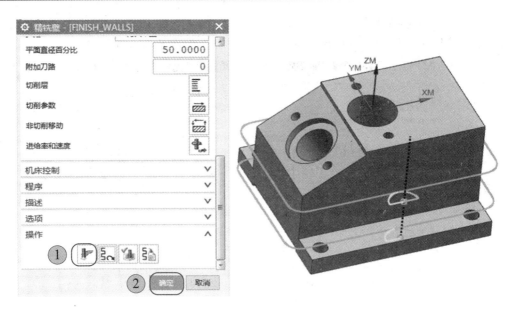

图 2-61　生成【精铣壁】加工程序

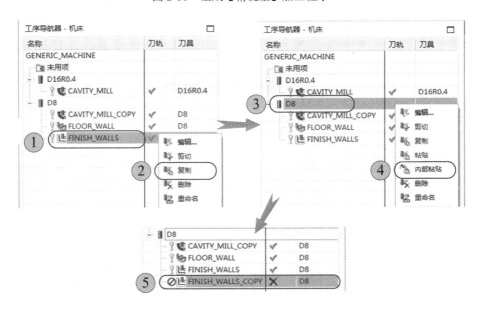

图 2-62　复制出新的【精铣壁】程序

(10) 双击打开新复制的【精铣壁】对话框，单击 📦【指定部件边界】图标，打开【部件边界】对话框。首先单击列表中 ✖【移除】按钮，删除列表中所有的边界线，然后在【边界】的【选择方法】选项中选择【曲线】的方式创建加工边界，设置【边界类型】为【封闭】，【刀具侧】为【外侧】，【平面】选择【指定】。单击工件下凸台表面为加工起始平面，如图 2-63 圆圈 5 所示，然后手动选择轮廓外圈线在指定的平面生成新的加工边界，如图 2-63 圆圈 8 所示，单击【确定】退出设置，具体操作步骤如图 2-63 所示。

(11) 单击 🔲【指定底面】图标，弹出【平面】对话框。设置加工底面，选择工件底面为加工底面，如图 2-64 所示。

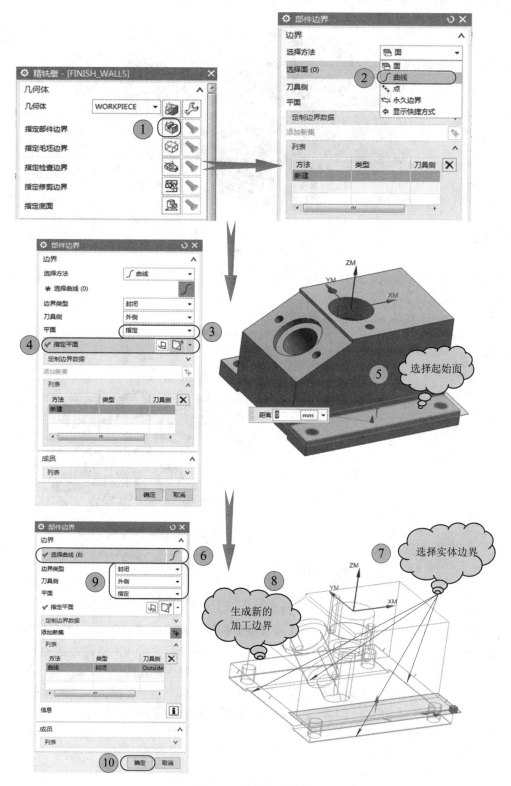

图 2-63　设置部件边界

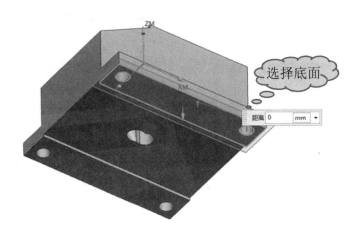

图 2-64　选择加工底面

(12) 单击 【生成】图标，计算加工程序，生成精加工底面侧壁路径，单击【确定】按钮退出【精铣壁】设置，如图 2-65 所示。

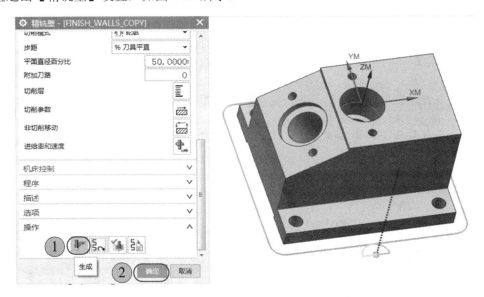

图 2-65　生成【精铣壁】加工程序

(13) 用相同的方法分别复制三个【精铣壁】程序，完成中间三个内孔的精加工程序。下面我们只介绍其中一个圆轮廓线的选取方法，其余两个圆轮廓线的选取方法参照第一个圆的方法。以下只介绍程序中不同的地方，其他参数都默认原先的程序。

(14) 双击打开新复制的精铣壁程序，单击 【指定部件边界】图标，打开【部件边界】对话框。首先单击列表中 【移除】按钮，删除所有上次已选择的边界线，然后在【边界】的【选择方法】选项中，选择【曲线】的方式创建加工边界，设置【边界类型】为【封闭】，【刀具侧】为【内侧】，【平面】为【指定】。单击工件上表面为加工起始平面，如图 2-66 圆圈 7 所示，然后手动选择轮廓外圈线，在指定平面生成新的加工边界，单击【确定】退出设置，具体操作步骤如图 2-66 所示。

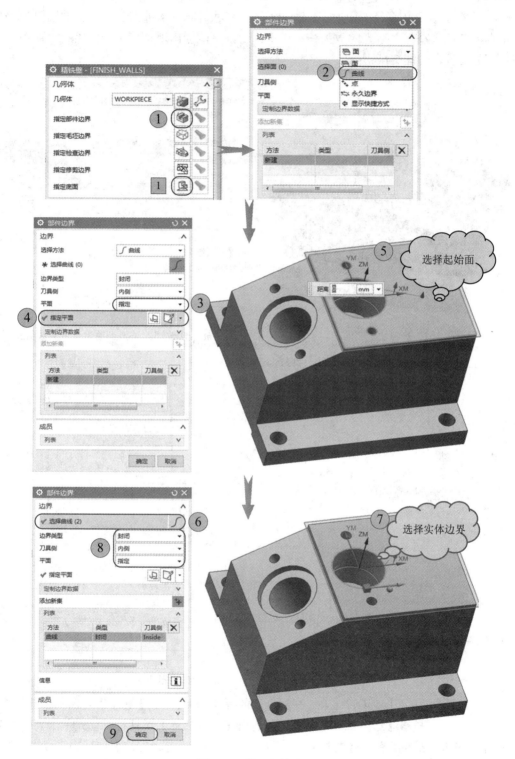

图 2-66 设置部件边界

(15) 单击 🔳【指定底面】图标，如图 2-66 方块 1 所示，弹出【平面】对话框。选择第一层圆的底面为加工底面，如图 2-67 所示。

(16) 在加工圆孔直径变小时，【非切削移动】对话框中【进刀】标签里的圆弧半径需要根据内圆的大小做出适当的放大缩小操作，否则可能会出现螺旋下刀的路径，而不是圆弧进刀的路径，如图 2-68 所示。

图 2-67　选择加工底面　　　　　图 2-68　修改圆弧进刀半径为合适的数值

(17) 分别生成三个内孔的精加工程序如图 2-69 所示。

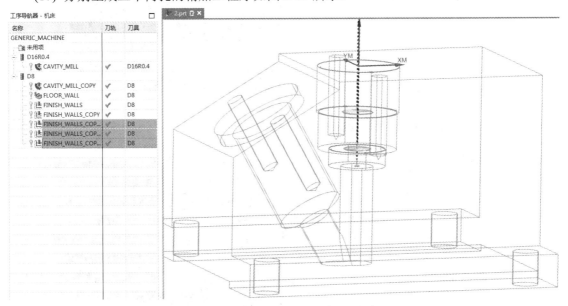

图 2-69　分别生成三个内孔的精加工程序

(18) 编辑精加工上侧面深 1.185 开放直边的程序，从【上面】程序组复制一个新的【精铣壁】程序。单击 ⬚【指定部件边界】图标打开【部件边界】对话框。首先单击列表中 ✖【移除】按钮，删除所有已选的边界线。然后在【边界】【选择方法】选项中，点击【选择曲线】的方式创建加工边界，设置【边界类型】为【开放】，【刀具侧】为【左】，【平面】为【指定】。单击工件上表面为加工起始平面，然后手动选择上面左侧边轮廓线为加工边界，单击【确定】退出设置，具体步骤如图 2-70 所示。

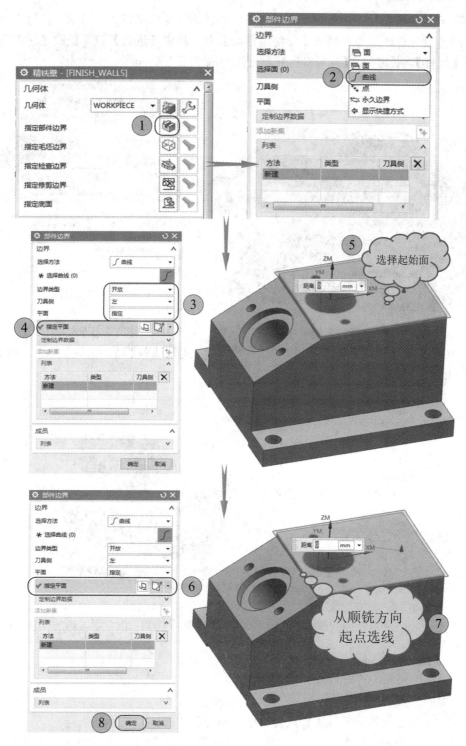

图 2-70　设置部件边界

(19) 单击 ⬚【指定底面】图标，弹出【平面】对话框。选择工件上表面，在【距离】对话框中输入"-1.185"使平面降到台阶底部形成加工底面，如图 2-71 所示。

(20) 开放轮廓精加工一般使用直线进刀的方法而不是圆弧进刀的方法，在这里我们将进刀方法修改为直线进刀。打开【非切削移动】对话框中【进刀】设置里的【进刀类型】选择为【线性】，长度为"【50%刀具】"。设置方法如图 2-72 所示。

图 2-71　选择加工底面　　　　　　　　图 2-72　修改进刀方式为直线进刀

(21) 单击 【生成】图标，计算加工程序。生成【精铣壁】精加工侧壁路径，单击【确定】按钮退出【精铣壁】设置，如图 2-73 所示。

图 2-73　生成精铣左侧壁加工程序

2.4.5　打孔程序编制

1.　创建 φ6 中心钻程序

创建 φ6 中心钻编制完成工件孔位的中心钻加工程序。

(1) 创建一个 φ6 的中心钻，名称设置为【中心钻 D6】，刀具号为 3 号。创建中心点孔程序，单击 【创建工序】图标，弹出【创建工序】对话框，在【类型】下拉菜单中选择【hole_making】孔加工选项，【工序子类型】选项中选择 【中心钻】加工方法，【程序】选项中选择【1】程序组，【刀具】选项中选择刚创建的【中心钻 D6】刀具，【几何体】选项中选择【1】，最后单击【确定】按钮打开【定心钻】对话框，具体操作步骤如图 2-74 所示。

（2）在【定心钻】对话框中单击 【指定特征几何体】图标，如图 2-75 圆圈 5 所示。由于中间圆孔上面几层的圆孔已经被加工完毕，打孔只需要加工最下层 φ5 的孔，所以起始平面不是从工件表面开始，而是从 φ5 圆表面开始，因此在选择孔位前要先把【过程工件】选项设置为【使用 3D】，这样软件就能自动识别出应从哪个位置开始打孔，然后选择正面需要点中心钻的所有孔位，按住键盘按键【Shift】，单击列表中最下面的孔和最上面的孔，松开【Shift】按键，这时列表中的所有孔都显示为蓝色背景状态。单击 图标选择 用户定义(U) 选项，【深度】数值输入 "1 mm" 为点孔深度，单击【确定】退出设置，具体操作步骤如图 2-76 所示。

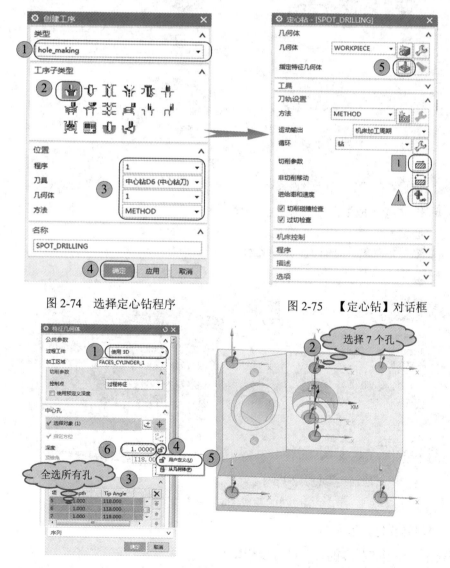

图 2-74　选择定心钻程序　　　　　图 2-75　【定心钻】对话框

图 2-76　选择孔位设置加工深度

(3) 单击 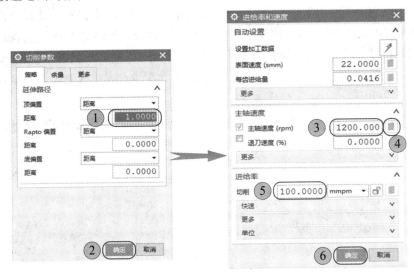 【切削参数】图标,如图 2-75 方框 1 所示,打开【切削参数】对话框,设置顶偏置【距离】为"1 mm",能减小打孔起始距离提高加工速度,然后单击【确定】退出对话框,如图 2-77 所示。

(4) 单击 【进给率和速度】图标,如图 2-75 三角 1 所示,设置【主轴速度】为"1200 rpm",单击右侧【计算器】图标,【进给率】的【切削】值为"100 mmpm",单击【确定】按钮退出对话框,如图 2-78 所示。

图 2-77　设置打孔起始距离　　图 2-78　设置转速和进给速度

(5) 单击 【生成】图标计算加工程序,生成【定心钻】加工路径,单击【确定】按钮退出【定心钻】设置,如图 2-79 所示。

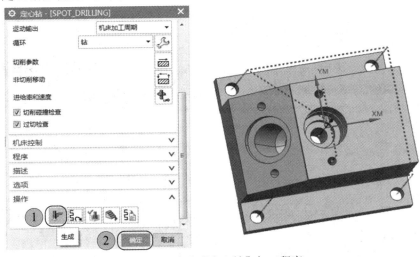

图 2-79　生成【定心钻】加工程序

2. 创建钻头打孔程序

创建第一序所有打孔的程序操作步骤如下:

(1) 单击【机床】视图,打开"创建工序"对话框,然后建立一把 φ2.5 的钻头,刀号为 4 号，刀具名称为【钻头 D2.5】。

(2) 创建钻孔程序，单击 【创建工序】图标，弹出【创建工序】对话框，在【类型】下拉菜单中选择【hole_making】孔加工选项，【工序子类型】选项中选择 【钻孔】加工方法，【程序】选项中选择【1】程序组，【刀具】选项中选择刚创建的【钻头 D2.5】刀具，【几何体】选项中选择【1】，最后单击【确定】按钮，如图 2-80 所示，打开【钻孔】对话框如图 2-81 所示。

(3) 在【钻孔】对话框中单击 【指定特征几何体】图标，如图 2-81 圆圈 4 所示。选择两个 φ2.5 直径的孔位，如图 2-82 所示。

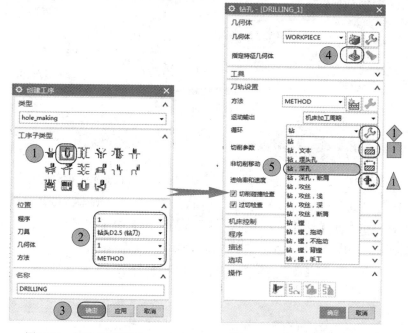

图 2-80　创建钻孔程序　　　　　图 2-81　选择孔位和啄钻方法

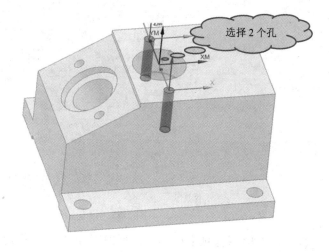

图 2-82　选择两个 φ2.5 直径孔位

(4) 单击 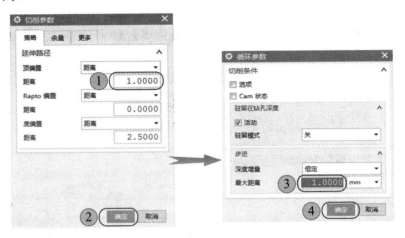【切削参数】图标，如图 2-81 方框 1 所示，设置顶偏置【距离】为"1 mm"，以减小打孔起始距离提高加工速度，然后单击【确定】退出对话框，如图 2-83 所示。

(5) 在循环下拉菜单下选择【钻，深孔】选项，设置为【啄钻加工】，如图 2-81 圆圈 5 所示，然后单击 【编辑循环】图标，如图 2-81 菱形 1 所示，设置【步进】【最大距离】为"1 mm"，此数值是啄钻加工中每层打孔的深度值，然后单击【确定】按钮退出对话框，如图 2-84 所示。

图 2-83　设置打孔起始距离　　　　图 2-84　设置啄钻每层的深度

(6) 单击 【进给率和速度】图标，如图 2-81 三角 1 所示，设置【主轴速度】为"1000 rpm"，单击右侧【计算器】图标，设置【进给率】中【切削】值为"100 mmpm"，然后单击【确定】按钮退出对话框，如图 2-85 所示。

图 2-85　设置转速和进给速度

(7) 单击 【生成】图标，计算加工程序，生成【钻孔】加工路径，单击【确定】按钮退出【钻孔】设置，如图 2-86 所示。

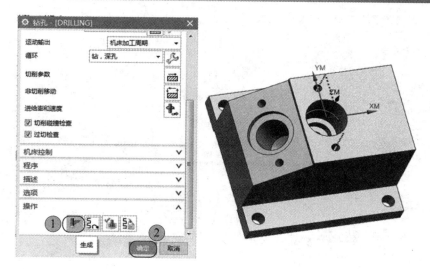

图 2-86　生成底孔加工程序

(8) 创建 φ4 钻头 5 号刀和 φ5 钻头 6 号刀，按照上面的方法完成 4 个 φ4 圆孔和中间一个 φ5 圆孔的加工程序。具体方法如图 2-87～图 2-90 所示。

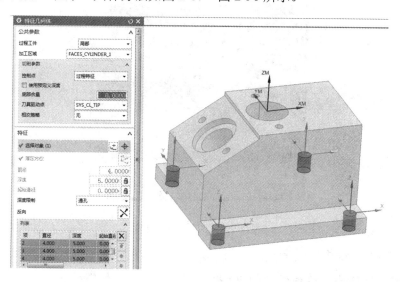

图 2-87　φ4 孔位选择

图 2-88　φ4 孔加工路径

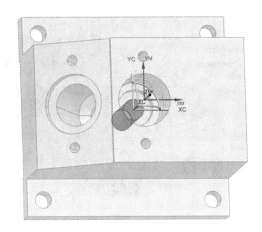

图 2-89 φ5 孔位选择

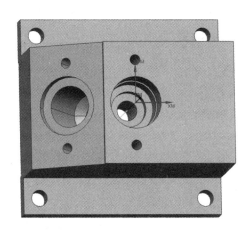

图 2-90 φ5 孔加工路径

3. 创建 M3 丝锥攻丝程序

创建 M3 丝锥编辑 M3 螺纹孔的攻丝程序具体操作步骤如下：

(1) 创建 M3 丝锥。单击 【机床】视图图标，如图 2-91 圆圈 1 所示，将【工序导航器】切换到【机床】视图页面。单击 【创建刀具】图标，弹出【创建刀具】对话框。

(2) 在【类型】下拉菜单中选择【hole_making】孔加工类型，如图 2-91 圆圈 3 所示。在【刀具子类型】选项中选择 【丝锥】图标，如图 2-91 圆圈 4 所示。在【名称】选项中输入刀具名称"M3"，如图 2-91 圆圈 5 所示，然后单击【确定】按钮，弹出【中心钻刀】对话框。

(3) 设置【颈部直径】为"3mm"，【螺距】为"0.5"，【刀具号】和【补偿寄存器】都输入"7"，其他参数无需设置，按默认值就可以，最后单击【确定】按钮退出设置，如图 2-92 所示。

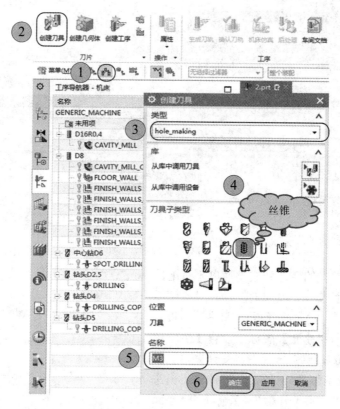

图 2-91　【创建刀具】对话框

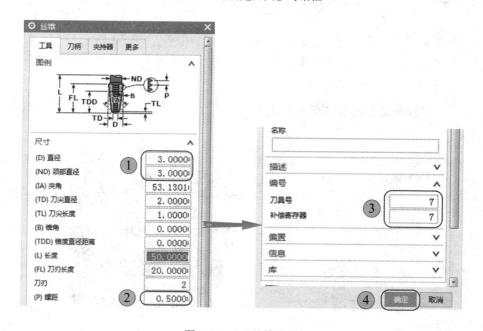

图 2-92　M3 丝锥设置

（4）复制 φ2.5 钻头打孔程序到 M3 丝锥下。双击打开钻孔程序，在【循环】下拉菜单中选择【钻，攻丝】。在弹出的【循环参数】对话框中单击【确定】按钮，如图 2-93 所示。

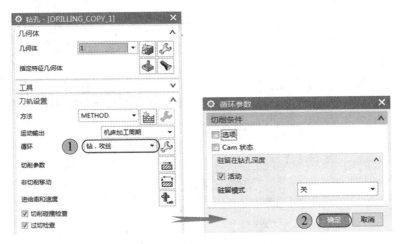

图 2-93　选择攻丝选项

(5) 丝锥攻丝时的转速和进给速度是有配比
关系的，不能随便指定，攻丝的转速和进给速度
的计算公式是 S(转速) $\times P$(螺距) $= F$(进给速度)。
单击 【进给率和速度】图标，设置【主轴速
度】为 "100 rpm"，单击转速右侧【计算器】图
标。已知 M3 螺距为 0.5，通过公式计算得出进
给速度为 50。【进给率】中【切削】选项输入 "50
mmpm"，单击【确定】按钮退出对话框，如图
2-94 所示。

(6) 单击 【生成】图标，计算加工程序，
生成【M3 丝锥】加工路径，单击【确定】按钮
退出【钻孔】设置，如图 2-95 所示。

图 2-94　设置转速和进给速度

图 2-95　生成 M3 螺纹加工程序

(7) 仿真编辑完的加工程序，单击左上方 【程序顺序】视图图标，在【程序顺序】
视图界面全选所有编完的加工程序，如图 2-96 所示。

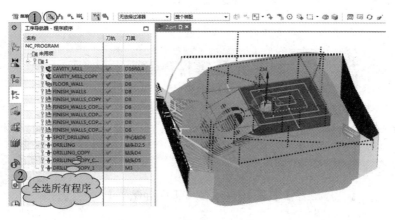

图 2-96　【程序顺序】视图全选所有加工程序

(8) 单击【主页】图标，如图 2-97 所示，打开【刀轨可视化】对话框。

图 2-97　单击【确认刀轨】图标

(9) 单击【3D 动态】标签，切换模拟动画为三维立体模型，然后单击下方▶【播放】图标，完成工件底面路径的加工仿真，如图 2-98 所示。

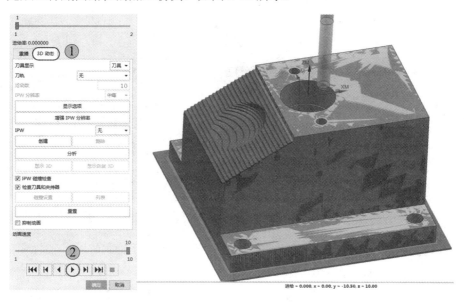

图 2-98　工序一加工路径仿真

2.5 批量加工零件的第二序加工

2.5.1 创建父节点组

1．创建程序组

光学医疗器械元件第二序和第
三序加工程序编制

首先单击 ▣ 【程序顺序】视图图标，使【工序导航器】
切换到【程序顺序】视图，然后单击屏幕左上角 ▣ 【创建程
序】图标，弹出【创建程序】对话框，【名称】输入"2"，单
击【确定】弹出【程序】对话框，再单击【确定】按键退出【程
序】对话框。在【程序顺序】视图导航器下增加了一个【2】的程序组，操作步骤如图 2-99
所示。

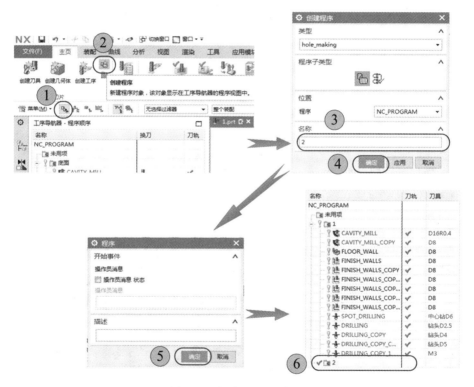

图 2-99　创建【2】程序组

2．创建坐标系和几何体

创建第二序坐标系和几何体的操作步骤如下：

(1) 将【工序导航器】切换到【几何】视图页面，右键单击【MCS_1】坐标系，选择
【复制】选项，然后右键单击【GEOMETRY】选择【内部粘贴】。这时把第一序的坐标系、

几何体和全部程序新复制一份，如图 2-100 圆圈 5 所示。

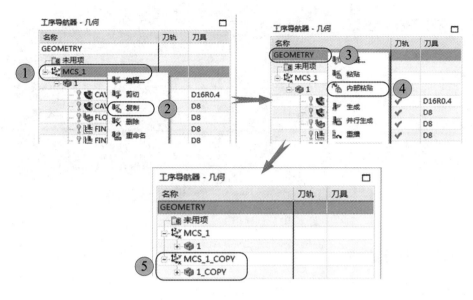

图 2-100　复制第一序的坐标系、几何体和程序

(2) 修改坐标系和几何体名称，右键单击【MCS_1_COPY】选择【重命名】选项，修改坐标系名称为"MCS_2"，然后右键单击【1_COPY】选择【重命名】选项，修改几何体名称为"2"，并删除新复制过来提示报错的所有程序，操作步骤如图 2-101 所示。

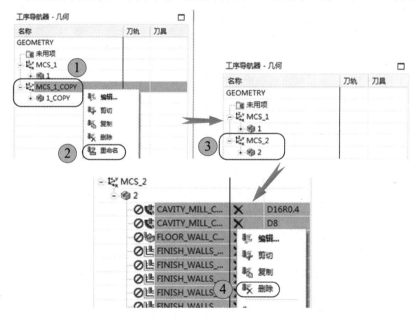

图 2-101　修改第二序坐标系和几何体名称

(3) 修改第二序坐标系位置，双击【MCS_2】图标，在弹出【MCS 铣削】对话框中选择 【坐标系】选项，在【坐标系】对话框中选择 动态 图标，按照图 2-102 位置选择坐标系方向，并把坐标系向下移动 11 mm 到工件最左面中间下端。

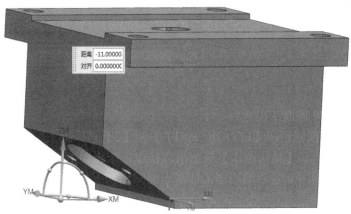

图 2-102　创建出第二序坐标系

2.5.2　粗加工程序编制

使用 φ16R0.4 钻铣刀粗加工零件第二序程序如下：

(1) 单击【程序顺序】视图制作第二序加工程序。复制【1】程序组的粗加工程序【CAVITY_MILL】到【2】程序组下，形成一个新的【型腔铣】程序。

(2) 双击打开刚复制的【CAVITY_MILL_COPY_1】图标，打开【型腔铣】对话框。在【几何体】选项中把【几何体】选择为新建立的【2】几何体，这时加工坐标系和几何体就会切换为第二序的坐标系和几何体，如图 2-103 所示。

图 2-103　选择第二序几何体

(3) 单击 【切削层】图标，设置第二序加工深度。由于第一序已经加工过正面大部分地方，在底面加工时只需要加工到底面凹槽即可。打开【切削层】对话框后，直接单击如图 2-104 所示平面，测得加工深度范围为 "1.5 mm"。单击【确定】按钮退出【切削层】对话框。

(4) 其余参数不需要修改，直接单击 【生成】图标计算出新的加工路径。单击【确定】退出【型腔铣】对话框，如图 2-105 所示。

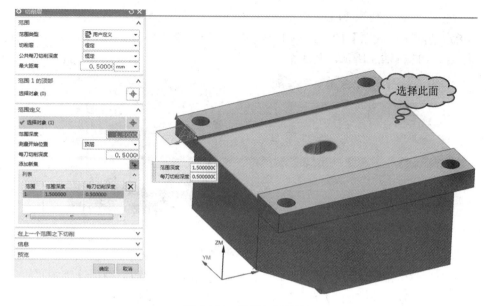

图 2-104　选择加工底面

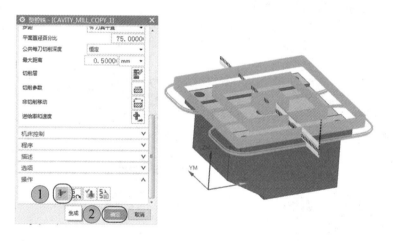

图 2-105　生成第二序粗加工程序

2.5.3　精加工底面程序编制

使用 φ8 合金刀完成工件底平面的精加工程序编制。

(1) 在【程序顺序】视图中制作第二序精加工底面程序。单击右键复制【1】程序组的精加工底面程序【FLOOR_WALL】到【2】程序组下，形成一个新的【底壁铣】程序，如图 2-106 圆圈 1 所示。

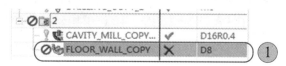

图 2-106　复制第一序的精加工底面程序到第二序程序组中

(2) 双击打开新复制的【FLOOR_WALL_COPY】程序，修改参数如图 2-107 所示。单击■【指定切削区域底面】图标，单击✕删除原先选择的加工底面，然后重新选择加工底面，如图 2-107 圆圈 3 所示，共 3 个。

图 2-107　重新选择加工底面

(3) 其余参数全部默认原程序不用修改，最后单击▶【生成】图标计算加工程序，生成【底壁铣】加工路径，单击【确定】按钮退出【底壁铣】设置，如图 2-108 所示。

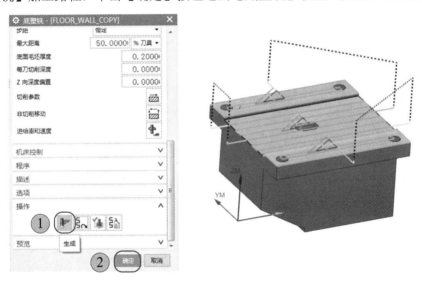

图 2-108　生成精加工底面程序

2.5.4　精加工侧面程序编制

编辑精加工第二序所有侧表面的加工程序方法如下：

(1) 单击【程序顺序】视图制作第二序精加工侧面程序。复制【2】程序组的精加工底面程序【FLOOR_WALL_COPY】到【2】程序组下，形成一个新的【底壁铣】程序，如图 2-109 所示。

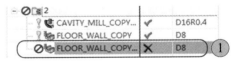

图 2-109　复制精加工底面程序到第二序程序组中

(2) 双击打开新复制的底壁铣程序。设置【切削模式】为【轮廓】的加工方法，如图 2-110 所示。

(3) 单击 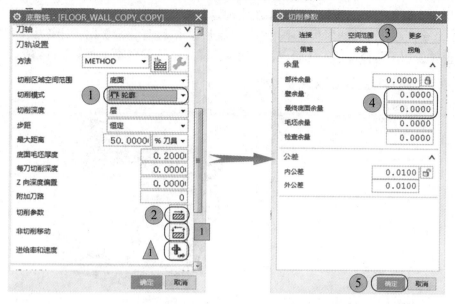 【切削参数】图标,如图 2-111 圆圈 2 所示,设置【余量】标签中【壁余量】为 "0 mm"。单击【确定】按钮退出【切削参数】对话框。

图 2-110　选择【轮廓】加工方法　　　　图 2-111　【壁余量】设置为 "0 mm"

(4) 单击  【非切削移动】图标,如图 2-110 方块 1 所示,设置【进刀】标签中的【进刀类型】为【线性】,【长度】为刀具直径的 "50%",提刀【高度】为 "1 mm"。单击【确定】退出设置,如图 2-112 圆圈 5 所示。

(5) 单击【更多】标签,打开刀具半径补偿功能,设置刀具补偿位置为【所有精加工刀路】,取消【最小移动】和【最小角度】中的数值,都改写为 "0",然后单击【确定】按钮退出【非切削移动】对话框,如图 2-113 所示。

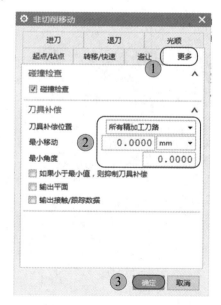

图 2-112　设置进刀方式为直线进刀　　　　图 2-113　打开刀具半径补偿功能

(6) 单击 【进给率和速度】图标，如图 2-110 三角 1 所示，修改【进给率】为 "500 mmpm"(精加工侧壁时走刀速度不宜过快，应控制在 500 mmpm 以下，否则加工表面粗糙度过大会影响加工质量，然后单击【确定】按钮退出【进给率和速度】对话框。

(7) 其余参数全部默认原程序不用修改，最后单击 【生成】图标计算加工程序，生成【底壁铣】加工路径，单击【确定】按钮退出【底壁铣】设置，如图 2-114 所示。

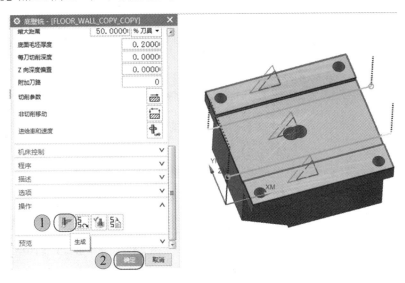

图 2-114　生成精加工侧面程序

(8) 仿真编辑第二序加工程序，单击左上方 【程序顺序】视图图标，在【程序顺序】视图界面全选所有编完的加工程序。单击【主页】下【确认刀轨】图标，打开【刀轨可视化】对话框。单击【3D 动态】，切换模拟动画为三维立体模型，然后单击下方 【播放】图标，完成工件底面路径的加工仿真，如图 2-115 所示。

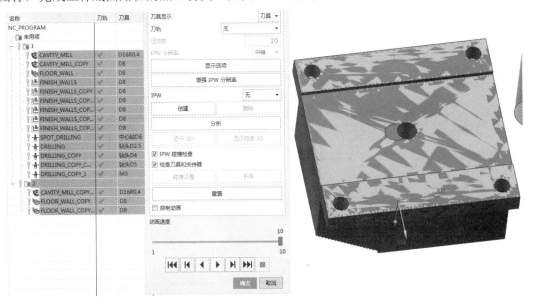

图 2-115　第二序编完程序的加工仿真

2.6　批量加工零件的第三序加工

2.6.1　创建父节点组

1. 创建程序组

首先单击 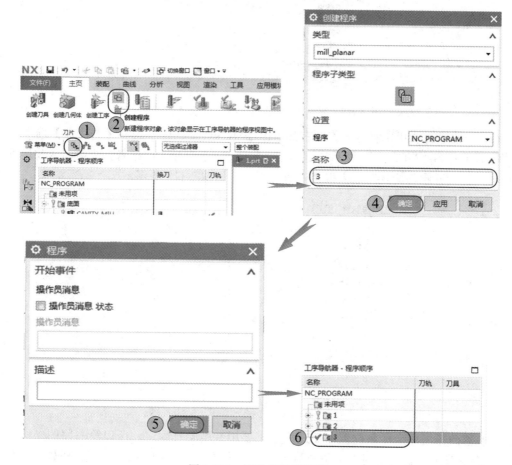【程序顺序】视图图标，使【工序导航器】切换到【程序顺序】视图，然后单击屏幕左上角 【创建程序】图标，弹出【创建程序】对话框，输入名称"3"，单击【确定】弹出【程序】对话框，再单击【确定】按键退出【程序】对话框。在【程序顺序】视图导航器下增加一个【3】的程序组，操作步骤如图 2-116 所示。

图 2-116　创建【3】程序组

2. 创建坐标系和几何体

创建第三序坐标系和几何体的操作步骤如下：

(1) 将【工序导航器】切换到【几何】视图页面，右键单击【MCS_1】坐标系，选择

【复制】选项，然后右键单击【GEOMETRY】选择【内部粘贴】。这时把上一步的坐标系、几何体和全部程序新复制一份，如图 2-117 所示。

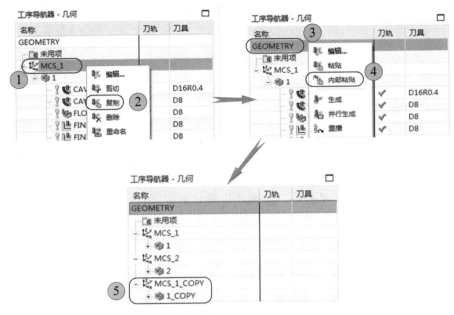

图 2-117　复制第一序的坐标系、几何体和程序

(2) 修改坐标系和几何体名称，右键单击【MCS_1_COPY】选择【重命名】选项，修改坐标系名称为"MCS_3"，然后右键单击【1_COPY】选择【重命名】选项，修改几何体名称为"3"，并删除新复制过来报错的所有程序，如图 2-118 所示。

图 2-118　修改第三序坐标系和几何体名称

(3) 修改第三序坐标系位置，第三序加工要用专用胎具固定，所以第三序的坐标位置应该放在胎具上。我们先调出隐藏的胎具，单击屏幕上方显示图标。单击选择显示两个虎钳钳口胎具，如图 2-119 所示。双击【MCS_3】图标，在弹出的【MCS 铣削】对话框中选择【坐标系】对话框，在【坐标系】对话框中选择 动态 图标，按照图 2-120 坐标系摆放的位置设置好工序三的加工坐标系。

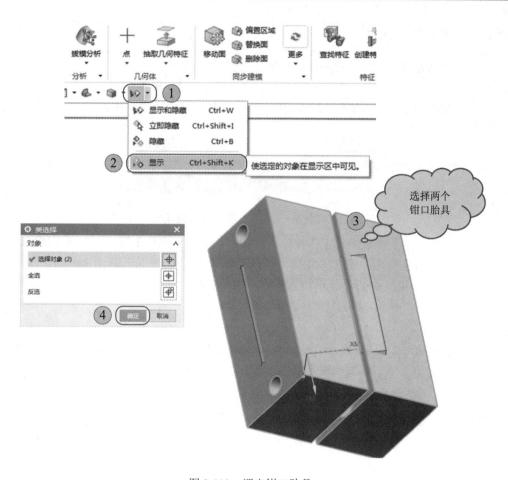

图 2-119　调出钳口胎具

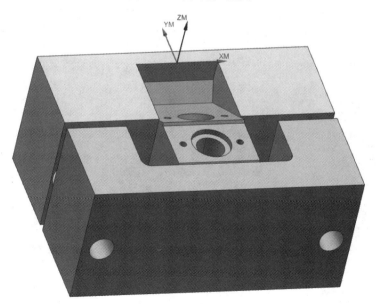

图 2-120　创建出第三序坐标系

2.6.2 粗加工程序编制

使用 φ8 铣刀粗加工零件第三序程序如下：

(1) 单击【程序顺序】视图制作第三序加工程序。复制【1】程序组中 φ8 铣刀粗加工程序【CAVITY_MILL_COPY】到【3】程序组下，形成一个新的【型腔铣】程序，如图 2-121 所示。

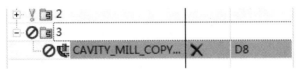

图 2-121　复制 φ8 刀粗加工程序到【3】程序组下

(2) 双击打开刚复制过来的【CAVITY_MILL_COPY_COPY】程序，打开【型腔铣】对话框。在【几何体】选项中把加工几何体选择为新建立的【3】几何体，这时加工坐标系和几何体就会切换为第三序的坐标系和几何体，如图 2-122 所示。

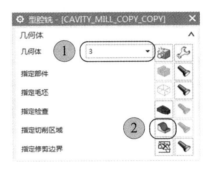

图 2-122　选择第三序几何体

(3) 单击【指定切削区域】图标，如图 2-122 圆圈 2 所示，选中工件斜面台阶孔表面，单击【确定】退出【切削区域】对话框，如图 2-123 所示。

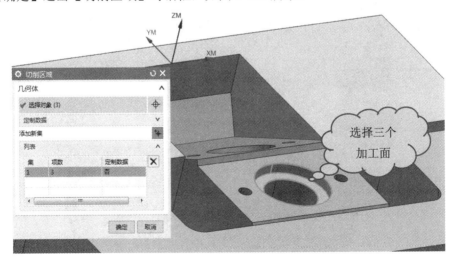

图 2-123　选择切削区域加工面

（4）单击 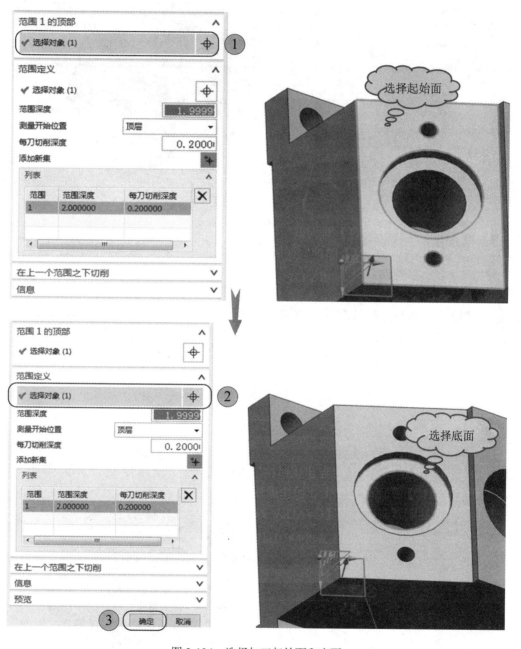【切削层】图标，设置第三序加工深度。由于第一序已经粗加工过斜面，因此在第三序粗加工时，只需加工上圆台阶孔即可。打开【切削层】对话框后直接单击如图 2-124 所示的起始平面和终止平面，测得加工范围深度为"2 mm"。单击【确定】选项退出【切削层】对话框。

图 2-124　选择加工起始面和底面

（5）其余参数都不需要修改，直接单击 【生成】图标，计算出新的加工路径。单击【确定】退出【型腔铣】对话框，如图 2-125 所示。

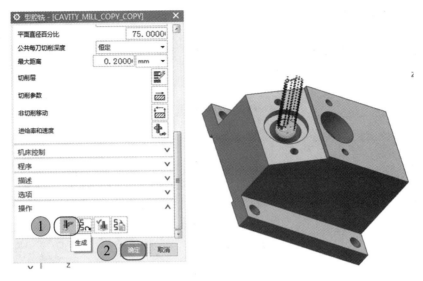

图 2-125 生成第三序粗加工程序

2.6.3 精加工底面程序编制

使用 φ8 合金刀完成工件底平面的精加工程序编制，步骤如下：

(1) 单击【程序顺序】视图制作第二序精加工底面程序。复制【1】程序组的精加工底面程序【FLOOR_WALL】到【3】程序组下，形成一个新的【底壁铣】程序，如图 2-126 所示。

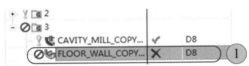

图 2-126 复制第一序的精加工底面程序到第三序程序组中

(2) 双击打开新复制的【FLOOR_WALL_COPY_1】程序，修改参数。单击 ⬛ 【指定切削区域底面】图标，单击 ✖ 删除原先选择的加工底面，重新选择加工底面，如图 2-127 所示。

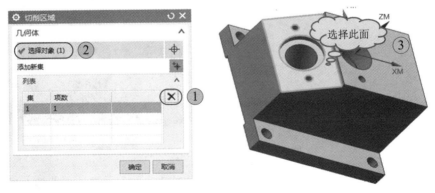

图 2-127 重新选择加工底面

（3）其余参数全部默认原程序不用修改，最后单击 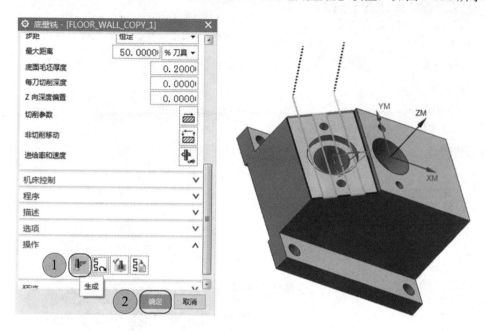【生成】图标，计算加工程序，生成【底壁铣】加工路径，单击【确定】按钮退出【底壁铣】设置，如图 2-128 所示。

图 2-128　生成精加工底面程序

2.6.4　精加工侧面程序编制

编辑精加工第三序所有侧表面的加工程序，方法如下：

（1）单击【程序顺序】视图制作第三序精加工圆台阶孔程序。复制【1】程序组的精加工孔的程序【FINISH_WALLS_COPY_COPY】到【3】程序组下，形成一个新的【精加工壁】程序，如图 2-129 所示。

图 2-129　复制出新的精加工壁的程序

（2）最上层圆的加工边界选择如下：双击打开新的精铣壁程序，在【几何体】下拉菜单中选择【3】。单击 【指定部件边界】图标打开【部件边界】对话框。首先单击列表中 ✖【移除】按钮删除所有已选择的边界线，然后在【边界】【选择方法】选项中选择【曲线】的方式创建加工边界，设置【边界类型】为【封闭】，【刀具侧】为【内侧】，【平面】为【指定】。左键单击工件上表面为加工起始平面，然后手动选择轮廓外圈线，在指定平面生成新的加工边界，单击【确定】退出设置，如图 2-130 所示。

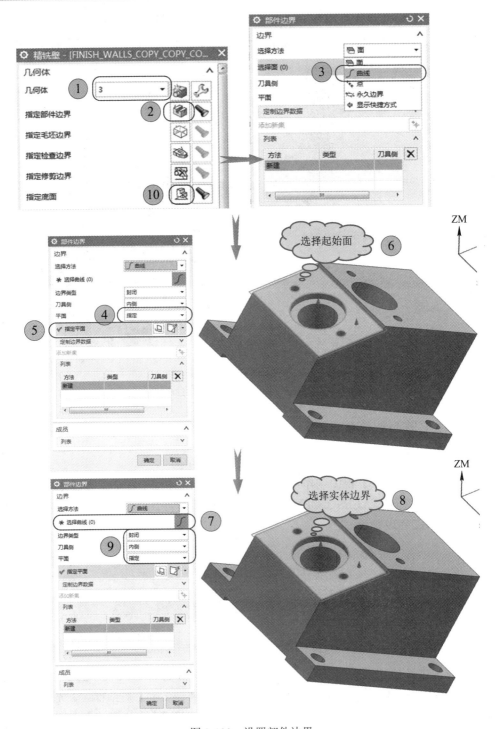

图 2-130　设置部件边界

　　(3) 单击 【指定底面】图标，如图 2-130 圆圈 10 所示，弹出【平面】对话框。设置加工底面时，选择第一层圆的底面为加工底面，如图 2-131 所示。

　　(4) 生成精加工圆台阶孔的程序如图 2-132 所示。

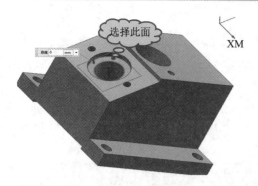

图 2-131　选择加工底面

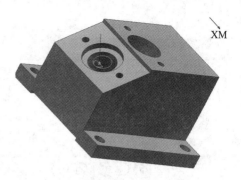

图 2-132　分别生成三个内孔的精加工程序

2.6.5　打孔程序编制

1．中心钻程序的编制

创建 φ6 中心钻加工程序。

(1) 单击【程序顺序】视图制作第三序点孔程序。复制【1】程序组中心钻点孔的程序【SPOT_DRILLING】到【3】程序组下，形成一个新的【定心钻】程序，如图 2-133 所示。

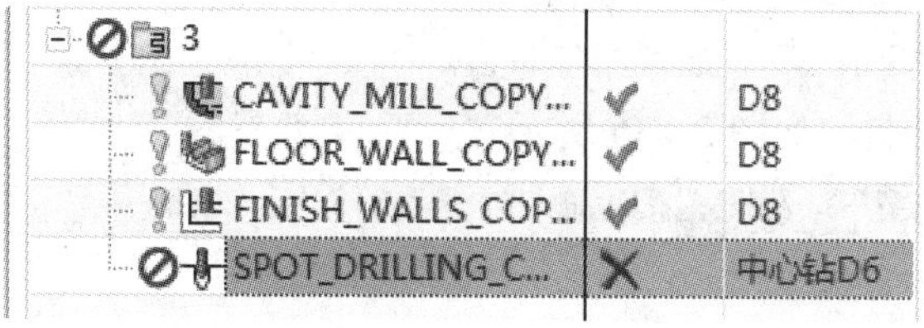

图 2-133　【定心钻】程序到【3】程序组里

(2) 双击打开新复制的【SPOT_DRILLING_COPY】程序，打开【定心钻】对话框。在【几何体】选项中把加工几何体选择为新建立的【3】几何体，这时加工坐标系和几何体就会切换为第三序的坐标系和几何体，如图 2-134 所示。

图 2-134　选择第三序几何体

(3) 在【定心钻】对话框中单击 【指定特征几何体】图标，如图 2-134 圆圈 2 所示。首先单击 图标删除原先选择的所有孔，将【过程工件】下拉菜单设置为【无】。选择斜面上需要点中心钻的三个孔位，然后按住键盘按键【Shift】，单击列表中最下面和最上面的孔，松开【Shift】按键。这时列表中的所有孔都显示为蓝色背景状态。单击 图标选择 用户定义(U) 选项，【深度】数值输入 "1 mm" 为点孔深度，如图 2-135 所示，单击【确定】

退出设置。

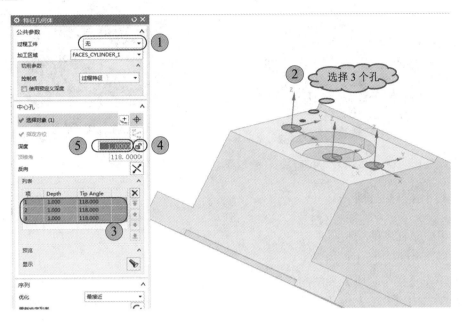

图 2-135　选择孔位设置加工深度

(4) 单击 ⤴【生成】图标,计算加工程序,生成【定心钻】加工路径,单击【确定】按钮退出【定心钻】设置,如图 2-136 所示。

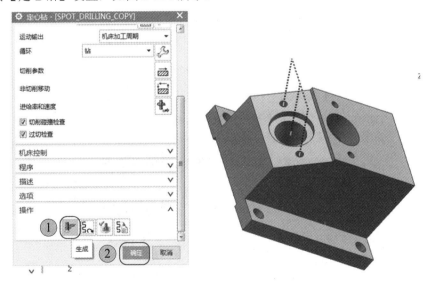

图 2-136　生成【定心钻】加工路径

2. 钻头打孔程序的创建

创建第三序所有打孔的程序操作步骤如下:

(1) 创建 φ10.2 钻头 8 号刀,并分别用 φ2.5 钻头、φ10.2 钻头、φ5 钻头按照第一序的打孔方法完成 2 个 φ2.5 圆孔,中间 1 个 φ10.2 盲孔和 1 个 φ5 通孔的程序。路径如图

2-137～图 2-139 所示。

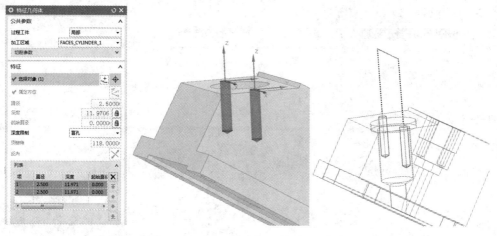

图 2-137　φ2.5 钻头加工斜面 M3 底孔

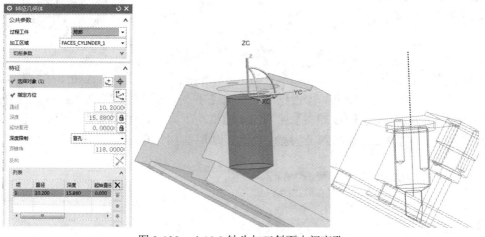

图 2-138　φ10.2 钻头加工斜面中间盲孔

（φ10.2 钻头转速应为 500 转）

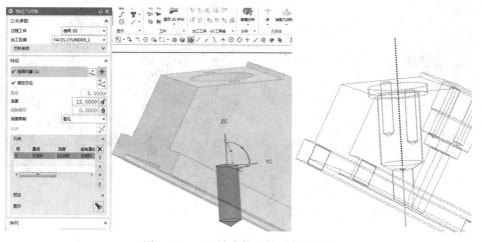

图 2-139　φ5 钻头加工斜面中间通孔

(2) 复制第一序 M3 丝锥攻丝程序到【3】程序组下，修改为第三序 M3 斜孔攻丝程序，如图 2-140 所示。

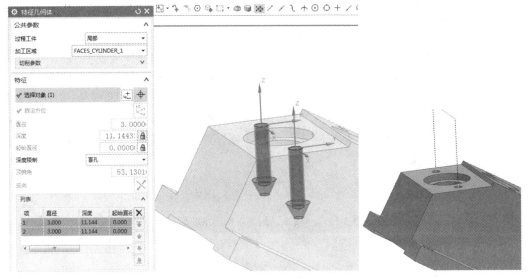

图 2-140　M3 丝锥斜孔攻丝程序

(3) 仿真编辑完的第三序加工程序，单击左上方 ⬚【程序顺序】视图图标，在【程序顺序】视图界面全选所有编完的加工程序。单击【主页】下【确认刀轨】图标，打开【刀轨可视化】对话框。单击【3D 动态】标签切换模拟动画为三维立体模型，然后单击下方 ▶【播放】图标，完成工件底面路径的加工仿真，如图 2-141 所示。

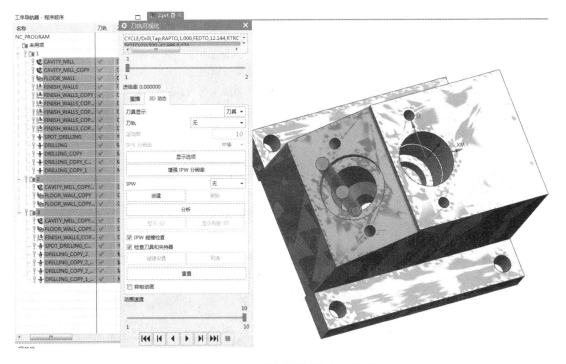

图 2-141　第三序编完程序的加工仿真

2.7　生成 G 代码文件

将编辑完的所有加工程序生成 G 代码文件。

(1) 电脑键盘按住【Ctrl】键，鼠标左键选择【1】程序组下所有程序，单击 【后处理】图标，如图 2-142 所示，弹出【后处理】对话框。

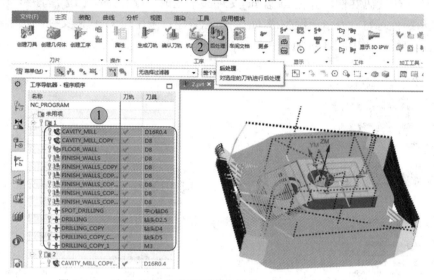

图 2-142　选择【1】程序组下所有程序，单击【后处理】图标

(2) 在【后处理器】选项中选择【FANUC0i】后处理文件，单击【输出文件】选项下的 【浏览以查找输出文件】图标，弹出【指定 NC 输出】对话框(首先在 D 盘创建 "nc" 文件夹)，选择 D:\nc 目录，输入文件名为 "1"，单击【OK】返回【后处理】对话框。确定文件名位置为 D:\nc\1，文件扩展名为.nc。单击【确定】退出设置，如图 2-143 所示。

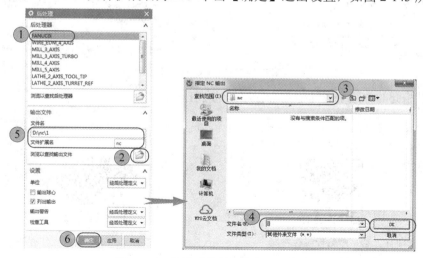

图 2-143　设置后处理文件位置及名称

(3) 单击【确定】后弹出【多重选择警告】对话框，单击【确定】按键将所有程序输出在一个程序组下显示，如图 2-144 所示。

图 2-144　弹出【多重选择警告】对话框

(4) 弹出 G 代码文件，并在 D 盘 nc 文件夹下生成 1.nc 文件。【1】程序组下所有程序前面都显示绿色对勾，表示已经生成 G 代码文件(注：未生成 G 代码文件的加工程序前显示为黄色感叹号)，如图 2-145 所示。

图 2-145　生成 G 代码文件

(5) 以相同的方法生成其他两个程序组的加工程序。

编程操作视频

项目二第一序编程操作视频　　　　　　　　　项目二第二、三序编程操作视频

课 后 练 习

　　按照本项目所学的知识完成课后练习中文件的程序编制，课后练习见图 2-146 和光盘"练习"文件夹中 2.prt 文件。

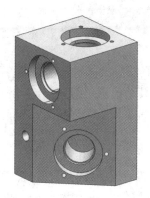

图 2-146　项目二课后练习图

项目三 自行车尾灯注塑模具动模型芯的程序编制

案例说明 ✍

本项目以自行车尾灯注塑模具动模型芯为案例,讲解自行车尾灯动模型芯的加工工艺、加工方法、切削刀具的选择以及曲面加工编程和钻孔加工的注意事项等。

学习目标 ✍

通过学习自行车尾灯注塑模具动模型芯的编程,读者应了解和掌握 NX 软件三轴曲面零件的加工编程方法(曲面区域轮廓铣、曲面清根铣等)以及钻孔加工程序的编制。做到举一反三,触类旁通。

学习任务 ✍

曲面类零件是各类模具加工中较为常用的零件,加工形状较为复杂,且大多为单件加工,常应用于注塑模具、冲压模具、压铸模具、吹塑模具以及复杂的曲面零件加工等。通过本案例可使读者掌握运用 NX 软件进行曲面区域轮廓铣、曲面清根铣加工的方法。

自行车尾灯注塑模具动模型芯三维图如图 3-1 所示。

图 3-1 自行车尾灯注塑模具动模型芯三维图

3.1 自行车尾灯注塑模具动模型芯的加工工艺规程

加工工艺规程是描述每步加工过程的,一般包括被加工的区域、加工类型(平面铣、曲面铣、孔加工等)、工序内容描述、零件装夹、所需刀具及完成加工所必需的其他信息。

自行车尾灯注塑模具动模型芯最大外形尺寸为 87 mm × 47 mm × 56.25 mm，材料为模具钢 P20，此产品单件加工。为便于铣削，在加工前对毛坯原料进行六面精加工铣磨，使其外形达到上述尺寸要求。

3.1.1　案例工艺分析

图 3-1 所示为自行车尾灯注塑模具动模型芯零件图，此工件需要上下两方向加工，且需要装夹两次零件完成工件的制作。为保证加工后零件的尺寸精度和形位公差，在编程前首先要确定好工件的加工工序方案。

工序一：

使用虎钳装夹先加工零件底部长方形挂台圆角以及钻孔部分。加工底部时由于长方形挂台高度为 6 mm，因此虎钳上表面到加工面的距离一定要大于 6 mm，否则可能会出现铣削到虎钳钳口的事故。安装毛坯后一定要测量工件的露出高度以确保安全生产。此次加工距离为 9.8 mm，加工时要注意加工部位的尺寸公差，按照图纸要求完成工件加工，如图 3-2 所示。

项目三第一序装夹方式

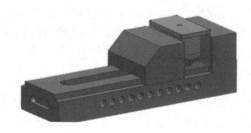

图 3-2　工序一安装示意图

工序二：

将虎钳装上等高垫铁，将第一序加工后的工件底面置于等高垫铁上，与其紧密贴合，然后用虎钳夹紧。确定工件露出钳口部分高度大于被加工零件高度的 50 mm，否则可能会出现铣削到虎钳钳口的事故，安装毛坯后一定要测量工件的露出高度以确保安全生产。加工时要注意加工部位的尺寸公差，按照图纸要求完成工件加工，如图 3-3 所示。

项目三第二序装夹方式

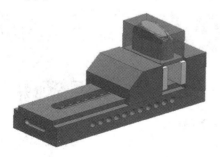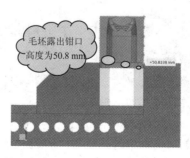

图 3-3　工序二安装示意图

3.1.2 案例加工刀具的选择

本案例自行车尾灯注塑模具动模型芯的加工材料为模具钢P20，最大外形尺寸为87 mm × 47 mm × 56.25 mm。最小凹圆角如图 3-4 所示为 R0.4527 mm，由于没有此尺寸的铣刀，故选用与其接近的 R1 球头硬质合金铣刀进行加工。型芯外轮廓由曲面与平面构成。首先对其进行粗加工，粗加工时要去除大量的材料，因加工时刀具承受的力会很大，所以要选取直径比较大的镶片钻铣刀 D16R0.4，这样加工时就不容易发生振刀、断刀、崩刃的现象。使用 R3 和 R2 球头硬质合金铣刀对加工材料分别进行二次和三次开粗，清理粗加工后残余的圆角部分余量，然后使用 R1 球头硬质合金铣刀φ16 精铣零件的轮廓和底面，最后按照零件孔位直径的大小选择合适的中心钻和钻头加工零件上所有的孔。

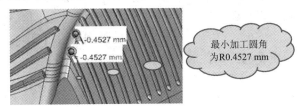

图 3-4　最小凹圆角示意图

按照上述分析的零件加工工艺方案和切削刀具的选择方式，合理安排零件的加工工艺过程。按照先粗后精、先面后孔、基准统一的原则设计本案例的加工工艺过程单，如表 3-1 所示。

表 3-1　自行车尾灯注塑模具动模型芯工艺过程单

工序号	顺序号	加工机床	工序内容	工序名	刀具名称
底面	1	加工中心	粗加工	型腔铣	D16R0.4
底面	2	加工中心	精加工侧壁	底壁铣	D16
底面	3	加工中心	打中心钻	定心钻	中心钻 D6
底面	4	加工中心	加工φ6.8 的孔	钻孔	钻头 D6.8
上面	5	加工中心	粗加工	型腔铣	D16R0.4
上面	6	加工中心	二次开粗	区域轮廓铣(区域铣削)	R3
上面	7	加工中心	清角加工	区域轮廓铣(清根)	R2
上面	8	加工中心	清角加工	区域轮廓铣(清根)	R1
上面	9	加工中心	底面与侧壁精加工	底壁铣	D16

3.2　打开模型文件进入加工模块

打开随书配套光盘，在"例题"文件夹中打开模型文件，并且进入加工模块。

(1) 启动 NX 12.0，单击左上角 📁【打开】按钮，在【打开】对话框中选择光盘"例题"文件夹中 3.prt 文件，如图 3-5 所示。

自行车尾灯动模型芯的程序编制

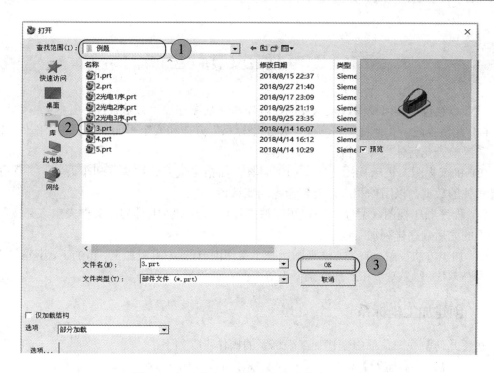

图 3-5　打开光盘"例题"文件夹中 3.prt 文件

(2) 单击【应用模块】选项，再单击【加工】图标(也可以直接单击键盘快捷键 Ctrl+Alt+M)启动 UG NX 12.0 "加工"模块，如图 3-6 所示。

(3) 在【加工环境】对话框中默认选择【CAM 会话配置】中的【cam_general】，在【要创建的 CAM 组装】选项中选择【mill_contour】曲面铣，如图 3-7 所示，单击【确定】按钮进入曲面铣加工界面。

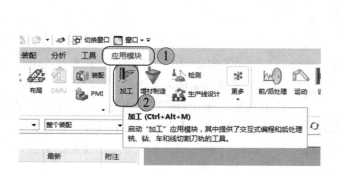

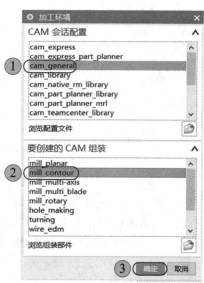

图 3-6　启动 UG NX12.0 "加工"模块　　　　图 3-7　【加工环境】对话框设置

3.3　建立父节点组

父节点组包括几何视图、机床视图、程序顺序视图和加工方法视图。

(1) 几何视图。几何视图可定义"加工坐标系"方向和安全平面，并设置"部件""毛坯"和"检查几何体"等参数。

(2) 机床视图。机床视图可定义切削刀具，如指定铣刀、钻头、和车刀等，并保存与刀具相关的数据，以用作相应后处理命令的默认值。

(3) 程序顺序视图。程序顺序视图能够把编好的程序按组排列在文件夹中，并按照从上到下的先后顺序排列加工程序。

(4) 加工方法视图。加工方法视图用来定义切削方法类型(粗加工、精加工、半精加工)，如"内公差""外公差"和"部件余量"等参数在此设置。

3.3.1　创建加工坐标系

在【几何】视图菜单中创建加工坐标系的操作步骤如下：

(1) 将【工序导航器】切换到【几何】视图页面，如图 3-8 所示。

(2) 右键单击【MCS_MILL】图标，选择【重命名】设置工序一坐标系名称为【MCS_1】，双击打开【MCS 铣削】对话框，如图 3-9 所示。

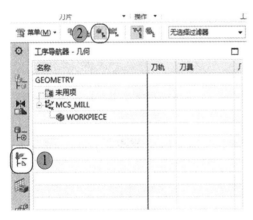

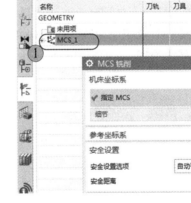

图 3-8　【几何】视图切换　　　　　　　　图 3-9　【MCS 铣削】对话框

(3) 由于加工原料为六面精磨料，为保证各面加工留量均匀，因此把坐标系放在毛坯中间位置。单击 [坐标系对话框] 图标，选择【对象的坐标系】选项，其功能是自动设置坐标为所选择平面的中心点位置。单击工件上表面方框自动捕捉工件的上表面中心点坐标位置。在安全平面没有干涉物的情况下可以选择默认状态【自动平面】，【安全距离】为"10 mm"。如有干涉物可把安全距离设置为"50 mm～100 mm"。最后单击【确定】按钮退出【MCS 铣削】对话框，坐标系设置方法如图 3-10 所示。

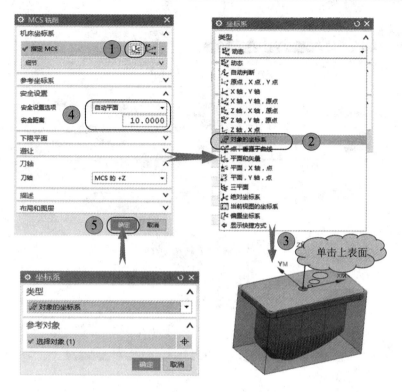

图 3-10 设置工序一加工坐标系

3.3.2 创建部件几何体

在【几何】视图菜单中创建加工部件几何体、零件毛坯、检查几何体的操作步骤如下：

(1) 右键单击图 3-8 所示【几何】视图中 ![WORKPIECE] 图标，选择【重命名】，设置工序一工件名称为"1"，双击打开【工件】对话框，如图 3-11 所示。

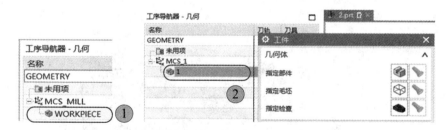

图 3-11 选择【WORKPIECE】图标弹出【工件】对话框

(2) 单击 ![icon]【选择或编辑部件几何体】图标，如图 3-12 圆圈 1 所示，弹出【部件几何体】对话框，单击被加工工件，使其成橘黄色，然后单击【确定】按钮退出【部件几何体】对话框，流程如图 3-12 所示。

(3) 单击 ![icon]【选择或编辑毛坯几何体】图标，如图 3-12 方框 1 所示。弹出【毛坯几何体】对话框，在【类型】选项中选择【包容块】选项。设置毛坯尺寸单边增加"0 mm"。单击【确定】按钮，返回【工件】对话框，然后再单击【确定】按钮退出【工件】对话框完成设置。设置方法如图 3-13 所示。

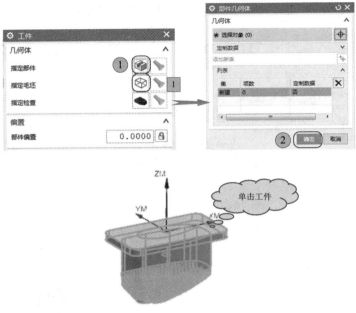

图 3-12　建立部件几何体

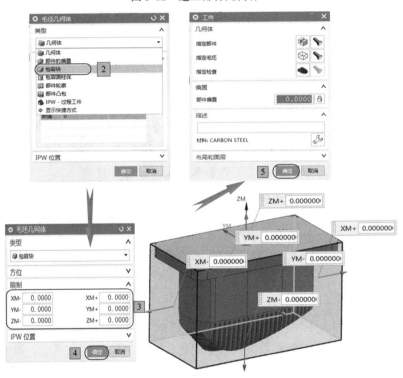

图 3-13　建立毛坯几何体

(4) 在选择完毛坯后，毛坯方框在编辑加工程序中再无用处，我们可以把它隐藏起来，以避免后序操作中产生误操作问题。具体方法为左键单击毛坯方框，按键盘快捷键 Ctrl + B 使其隐藏。如果想返回显示，则按键盘快捷键 Ctrl + Shift + K，然后左键选择要恢复的图形退出【显示】设置。

3.3.3　创建刀具

在【机床】视图下创建加工刀具步骤如下：

(1) 单击 【机床】视图图标，将【工序导航器】切换到【机床】视图页面，如图 3-14 所示。

(2) 单击【创建刀具】图标，如图 3-15 圆圈 2 所示，弹出【创建刀具】对话框。

图 3-14　【机床】视图　　　　　　　图 3-15　【创建刀具】对话框

(3) 选择 【平底刀】图标，创建平底刀。【名称】位置输入 "D16R0.4"（代表直径为 16 mm、圆角半径为 0.4 mm 的镶片钻铣刀），如图 3-16 所示。

(4) 【直径】设置为 "16"，【下半径】设置为 "0.4"，【刀具号】【补偿寄存器】【刀具补偿寄存器】三项均设置为 "1"（此数值代表刀具、刀具半径补偿和刀具长度补偿号，为避免发生撞机问题，最好设置为相同数字）。单击【确定】完成刀具创建，如图 3-17 所示。

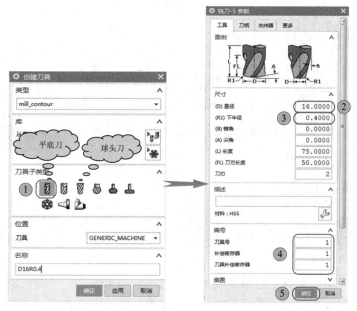

图 3-16　创建平底刀　　　　　　　图 3-17　平底刀设置

其他刀具在编辑加工程序前，按照给定参数自行设置。

3.3.4 创建程序组

在【程序顺序】视图中创建加工程序组文件夹，操作步骤如下：

(1) 将【工序导航器】切换到【程序顺序】视图页面，如图 3-18 所示。

图 3-18 【工序导航器】切换到【程序顺序】视图页面

(2) 双击【PROGRAM】程序组文字，如图 3-18 方框 1 所示，修改文件名为 "底面" (或使用右键单击【PROGRAM】程序组，选择【重命名】也可实现更改名称)，如图 3-19 所示。

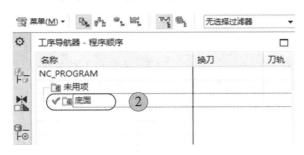

图 3-19 双击修改程序组名称

(3) 保存文件。

3.4 自行车尾灯注塑模具动模型芯的第一序加工

由前面的加工分析得出此工件的第一序加工，为零件底部朝上加工外形圆台。以下为第一序加工的程序编制方法。

3.4.1 粗加工程序编制

NX 软件在建立有模型图和毛坯的基础上编程时，使用最为简单有效的开粗程序就是曲面加工里的型腔铣，使用型腔铣可以完成绝大多数零件的开粗工作。在项目一中已经讲解了型腔铣的加工方法，由于此原料为六面精磨料，在粗加工时工件底面是不需要加工的，只需要精铣侧壁，因此下面介绍精铣壁的加工方法。为了更好地体现出加工程序的先后顺序，本案例在编程时全部使用【程序顺序】视图来完成程序的编制。

(1) 单击【创建工序】图标，如图 3-20 所示，弹出【创建工序】对话框。

图 3-20　单击【创建工序】图标

(2) 在【类型】下拉菜单中选择【mill_contour】曲面铣选项，如图 3-21 圆圈 2 所示。选择 【型腔铣】图标，如图 3-22 圆圈 3 所示，在【程序】下拉菜单中选择刚建好的【底面】程序组，【刀具】下拉菜单中选择【D16R0.4】的镶片钻铣刀，【几何体】下拉菜单中选择【1】几何体，如图 3-22 圆圈 4 所示，在【名称】栏中可以按照加工要求输入一个程序名称，本文在这里不做专门修改，按照默认名称填写，然后单击【确定】按钮，如图 3-22 圆圈 5 所示。

图 3-21　选择【mill_contour】曲面铣选项

图 3-22　创建【型腔铣】工序

(3) 在弹出的【型腔铣】对话框中，只要在【几何】视图中正确设定【WORKPIECE】即可。在进入型腔铣时，【指定部件】和【指定毛坯】选项应显示为灰色，右侧 🖉【显示】图标为彩色，单击可显示已选择的几何体部分。如果进入型腔铣后还能选择部件和毛坯，则说明【几何】视图中的【WORKPIECE】没有设定，或进入程序前没有选择几何体为【WORKPIECE】。型腔铣编程中一般不用设置【指定切削区域】，【指定检查】和【指定修剪边界】根据零件加工需求设置，在本例中不需要设置，如图 3-23 所示。

(4) 打开【工具】菜单下拉箭头，显示已选择的加工【刀具】为 "D16R0.4" 镶片钻铣刀(注：此项工作前序选择正确的情况下可忽略)。

(5) 打开【刀轴】菜单下拉箭头，显示出默认刀轴为【+ZM 轴】，三轴加工中心刀轴一般使用+ZM 轴，只有在使用多轴机床加工时才会修改此项，如图 3-24 所示(注：此选项三轴加工编程时不用选择，使用默认设置即可)。

图 3-23　几何体设置对话框

图 3-24　工具和刀轴选项

(6)【刀轨设置】为型腔铣参数设置的主要内容。【切削模式】选项中一般常用 跟随部件和 跟随周边 两种方式，【跟随部件】适合加工开放轮廓的工件，可以使刀具从外向内加工并从工件外下刀。【跟随周边】更适合加工封闭轮廓的工件，可以使刀具从内向外加工，减少型腔加工时的下刀位置变化。本案例底部开放轮廓加工，所以在【切削模式】中选择【跟随部件】的加工方法，如图 3-25 所示。

(7) 粗加工 XY 方向，刀具步距一般使用刀具直径的 70%～80%，精加工时步距使用刀具直径的 50%以下，80%～100%的 XY 方向步距一般情况下不推荐使用。首先，如果刀具步距太大，则每次切削都相当于满刀切削，刀具受力过大会影响刀具寿命和机床精度；其次，步距太大时加工完底面的光洁度很低，会出现接刀痕。因此在【平面直径百分比】选项中设置数值为"75%"，如图 3-26 圆圈 2 所示。

(8) D16R0.4 钻铣刀粗加工时的 Z 方向步距一般情况下取值 0.3 mm～0.7 mm 每层。【最大距离】选项设置的就是刀具 Z 方向的每层步距量，因此设置中间数值"0.5 mm"每层，如图 3-26 圆圈 3 所示。

图 3-25　选择【跟随部件】

图 3-26　XYZ 方向步距量设置

(9) 单击 【切削参数】图标，如图 3-26 圆圈 4 所示，打开【切削参数】对话框。

(10) 在【策略】标签中设置【切削顺序】为【深度优先】，【深度优先】会按照不同区域分别由上往下加工，从而可以减少抬刀和过刀路径，减少加工时间。【层优先】会按照同一深度在不同区域跳刀加工，从而会增加很多抬刀路径，增加加工时间。一般情况下优先选择【深度优先】，如图 3-27 所示。

图 3-27　【策略】标签设置

(11) 选择【余量】标签，设置【部件侧面余量】参数为"0.2 mm"(注：粗加工刀具余量一般设置为 0.2mm)，单击【确定】按钮退出【切削参数】对话框，如图 3-28 所示。

(12) 单击 【非切削移动】图标，如图 3-29 圆圈 1 所示，打开【非切削移动】对话框。

(13) 设置进刀参数时，首先设置【封闭区域】下刀参数，型腔内【进刀类型】一般采用【螺旋】下刀的方式。螺旋下刀加工原理是，在型腔尺寸能够满足螺旋下刀时，选用螺旋下刀的方式进刀。在型腔尺寸不能满足螺旋下刀时，选用斜线下刀或直线下刀的方式进行。螺旋下刀【直径】选用刀具直径的"50%"，【斜坡角度】设置为"5"，【高度】设置为"1 mm"，【最小斜坡长度】设置为"0"(注：加工刀具能够直线下刀时可以输入数字为"0"。如果加工刀具不能直线下刀，则此数字最小不能小于50，否则会发生撞机事故)。【开放区域】设置进刀【长度】为刀具的"50%"，抬刀【高度】设置为"1 mm"，以减少抬刀距离。具体参数设置如图 3-30 圆圈 2 和圆圈 3 所示。

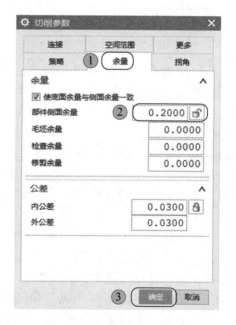

图 3-28　【余量】标签设置

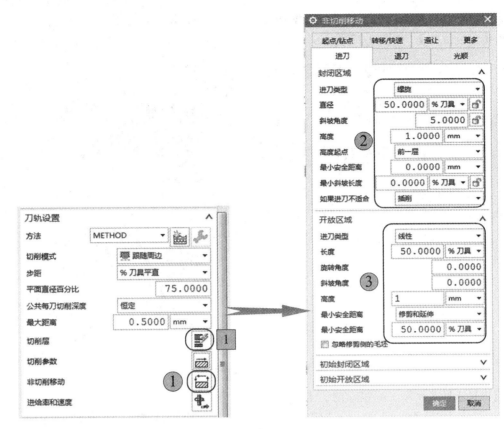

图 3-29　选择非切削移动　　　　图 3-30　进刀参数设置

(14) 单击【非切削移动】对话框【转移/快速】标签设置快速抬刀高度。为提高加工速度减少抬刀高度，把【区域之间】和【区域内】的【转移类型】都改为【前一平面】，并且把【安全距离】都设置为"1 mm"。但大部分快速移刀都是在工件零平面以下进行，所以要求机床的 G00 运动必须是两点间的直线运动，不能是两点间的折线运动，否则会发生撞机事故。加工前一定要在 MDI 下输入"G00 走斜线观察机床"的移动方式，如果不对，则需要修改机床参数。或者按照 NX12.0【转移/快速】的初始设置方式，把【转移类型】全部设置为【安全距离-刀轴】。每次提刀都回到安全平面，增加加工时间。最后单击【确定】按钮退出【非切削移动】对话框，具体参数设置如图 3-31 所示。

(15) 单击 【切削层】图标，如图 3-29 方框 1 所示，进入【切削层】对话框。

(16) 在【切削层】对话框中单击 多次，将列表中数值全部删除，如图 3-32 所示。

(17) 单击【范围定义】里的【选择对象】，选择零件如图 3-33 所示位置。测得图纸实际加工【范围深度】为"6 mm"，但是为了加工正面时都能加工到位，不留下接痕，故向下多加工 1 mm，所以【范围深度】值设置为"7 mm"，单击【确定】退出【切削层】对话框。

(18) 单击 【进给率和速度】图标，打开【进给率和速度】对话框。设置【主轴速度】为"2500 rpm"(注：输入"2500"后一定要按后面的【计算器】图标，否则会报警)。输入【进给率】【切削】为"1500 mmpm"，【更多】里【进刀】为"70%切削"，然后单击【确定】退出【进给率和速度】对话框，如图 3-34 所示。

图 3-31 【转移/快速】标签设置

图 3-32 删除列表数值

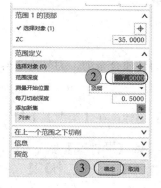

图 3-33 选择切削深度

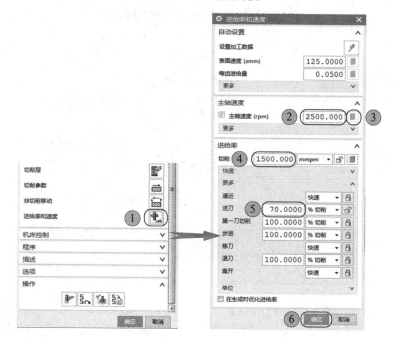

图 3-34 进给率和速度设置

(19) 单击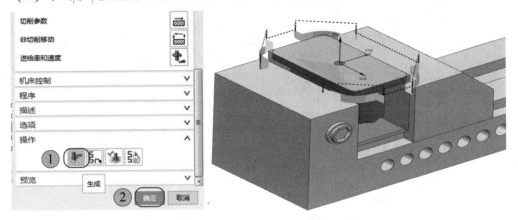【生成】图标计算加工路径。单击【确定】退出型腔铣,如图 3-35 所示。

图 3-35　生成刀具路径图

3.4.2　精铣壁程序编制

(1) 点开【机床】视图,建立一把直径为 16 mm 的硬质合金刀,刀号为 2 号,刀具名称为 D16。

(2) 创建精加工侧壁程序,单击　【创建工序】图标,在【类型】下拉菜单中选择【mill_planar】平面铣选项,【工序子类型】选项中选择　【精铣壁】加工方法,【程序】选项中选择【底面】程序组,【刀具】选项中选择刚创建的【D16 钻铣刀】,【几何体】选项中选择【1】,如图 3-36 圆圈 3 所示,最后单击【确定】按钮打开【精铣壁】对话框。

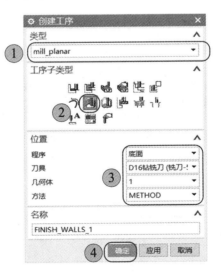

图 3-36　进入精铣壁加工

(3) 单击　【选择或编辑部件边界】图标,如图 3-37 圆圈 1 所示,弹出【部件边界】对话框,单击【选择曲线】,其他参数设置如图 3-37 圆圈 3 所示,最后单击【确定】。

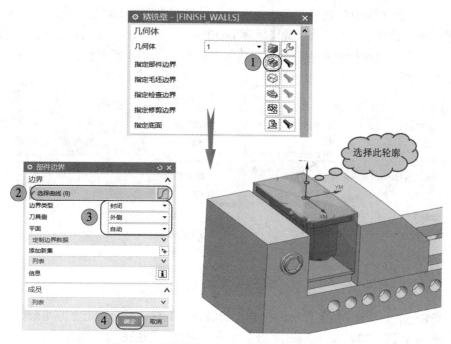

图 3-37　【部件边界】对话框选择工件轮廓线

（4）单击 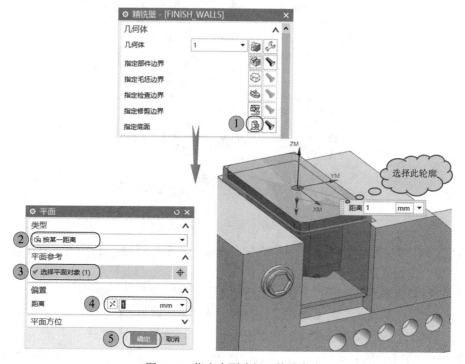 【选择或编辑底平面几何体】图标，弹出【平面】对话框，在【类型】下拉菜单中选择【按某一距离】，在【平面参考】下拉菜单中选择【选择平面】，在【偏置】【距离】里输入数值"1"，最后单击【确定】，如图 3-38 所示。

图 3-38　指定底面选择工件轮廓线

(5) 因为被加工表面只是一圈封闭轮廓的侧壁没有底面，所以更适合使用【轮廓】的加工方法，所以设置【切削模式】为【轮廓】，精加工时 XY 方向的步距量应小于刀具直径的 50%，在软件中【平面直径百分比】默认为"50%"，所以不需要修改，如图 3-39 所示。

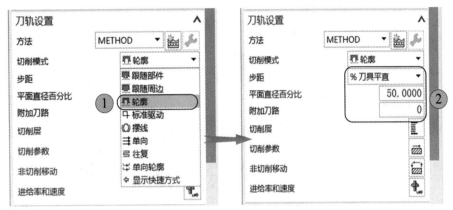

图 3-39 【刀轨设置】对话框设置

(6) 单击 💿【切削参数】图标，打开【切削参数】对话框，单击上方【余量】标签，设置【部件余量】为"0 mm"(注意：由于此程序是精加工侧壁程序，无需留余量，所以【部件余量】应为"0 mm"。)。减小公差的数值提高加工精度，设置【内公差】和【外公差】为"0.01mm"，单击【确定】按钮退出【切削参数】对话框，如图 3-40 所示。

图 3-40 加工余量设置

(7) 在【非切削移动】对话框中单击左上方【进刀】与【退刀】标签，在【进刀】与【退刀】标签中设置进刀与退刀参数，如图 3-41 所示。

(8) 单击 🔩【进给率和速度】图标，打开【进给率和速度】对话框，设置【主轴速度】为"2500 rpm"，单击右侧 🖩【计算器】图标。设置【进给率】【切削】为"500 mmpm"，然后单击【确定】按钮退出【进给率和速度】对话框，如图 3-42 所示。

(9) 单击 ▶【生成】图标计算加工程序，生成【精铣壁】加工路径，单击【确定】按钮退出【精铣壁】设置，如图 3-43 所示。

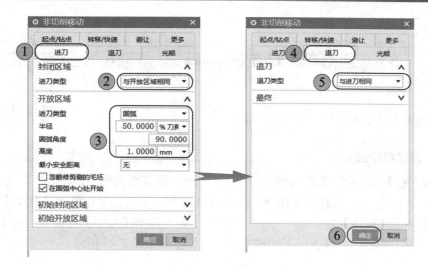

图 3-41　进刀与退刀参数设置

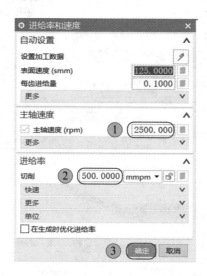

图 3-42　【进给率和速度】对话框

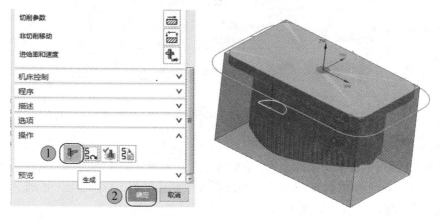

图 3-43　生成【精铣壁】加工路径

3.5 自行车尾灯注塑模具动模型芯的第二序加工

3.5.1 创建父节点组

1. 创建程序组

单击 【程序顺序】视图图标，使【工序导航器】切换到【程序顺序】视图，然后单击屏幕左上角【创建程序】图标，弹出【创建程序】对话框，输入名称"上面"，单击【确定】弹出【程序】对话框，再单击【确定】按键退出【程序】对话框。在【程序顺序】视图导航器下增加一个【上面】的程序组，操作步骤如图 3-44 所示。

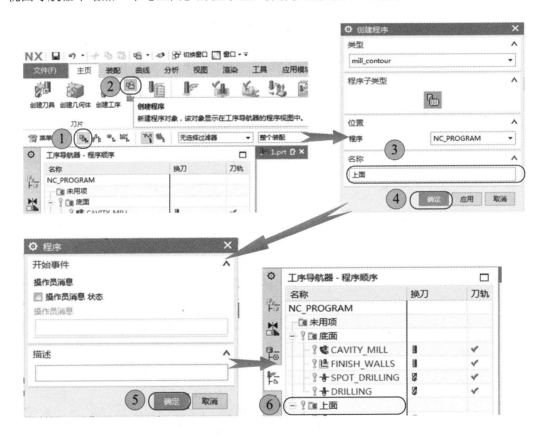

图 3-44　创建【上面】程序组

2. 创建坐标系和几何体

(1) 将【工序导航器】切换到【几何】视图页面，右键单击【MCS_1】坐标系选择【复制】选项，然后右键单击【GEOMETRY】选择【内部粘贴】。这时把上一步的坐标系、几何体和程序全部复制一份，如图 3-45 所示。

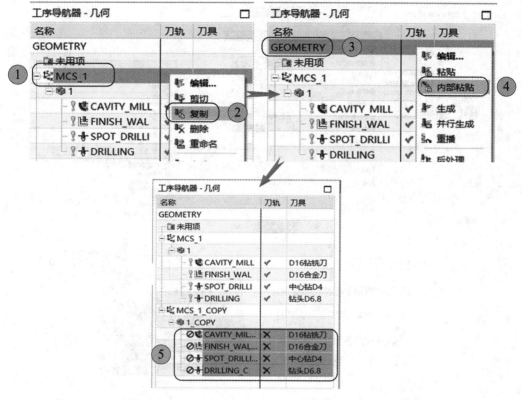

图 3-45　复制第一序的坐标系、几何体和程序

（2）修改坐标系和几何体名称，右键单击【MCS_1_COPY】选择【重命名】选项，修改坐标系名称为"MCS_2"，然后右键单击【1_COPY】选择【重命名】选项，修改几何体名称为"2"，如图 3-46 所示。

（3）按住键盘【Ctrl】键，选中【2】几何体中所有提示错误的程序，右键选择【删除】，删除所有复制过来的程序，如图 3-47 所示。

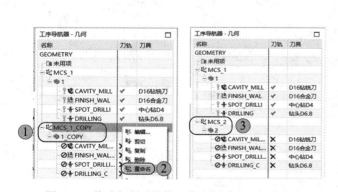

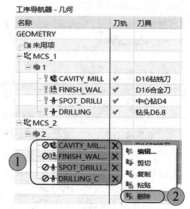

图 3-46　修改复制后的第一序坐标系和几何体名称　　　　图 3-47　删除复制的程序

（4）修改坐标系 Z 轴方向，双击 MCS_2 图标，在弹出的【MCS 铣削】对话框中选择【坐标系对话框】图标，再选择【对象的坐标系】，如图 3-48 所示。

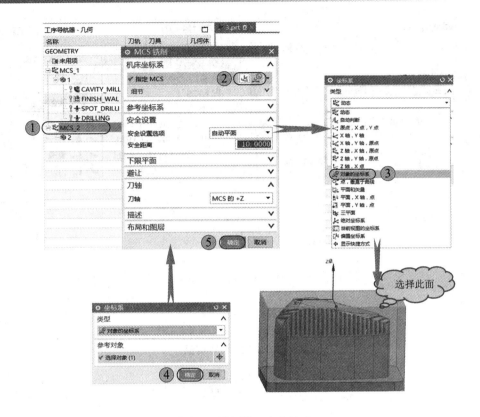

图 3-48　设置第二序坐标系

3.5.2　粗加工程序编制

(1) 单击【程序顺序】视图制作第二序加工程序。复制【底面】程序组的粗加工程序【型腔铣】到【上面】程序组下，形成一个新的【型腔铣程序】，如图 3-49 所示。

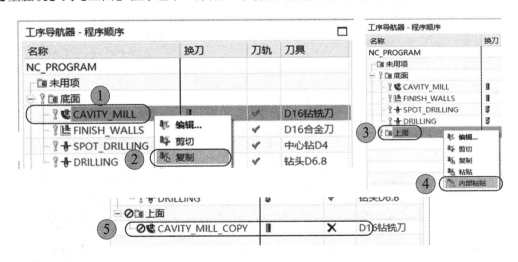

图 3-49　复制【型腔铣】程序到【上面】程序组里

(2) 双击打开刚复制过来的 CAVITY_MILL_COPY【型腔铣】图标，如图 3-49 圆圈 5

所示，打开【型腔铣】对话框。在【几何体】选项中把加工几何体选择为新建立的【2】几何体，这时加工坐标系和几何体就会切换为第二序的坐标系和几何体，如图 3-50 所示。

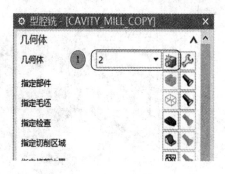

图 3-50　选择第二序几何体

(3) 单击 ![] 【切削层】图标设置第二序加工深度，由于第一序已经加工过底面挂台部分地方，因此在外框加工只需要加工到两个凸起台的底面即可。打开【切削层】对话框后直接单击如图 3-51 所示平面，测得加工【范围深度】为"50.2515mm"。单击【确定】退出【切削层】对话框。

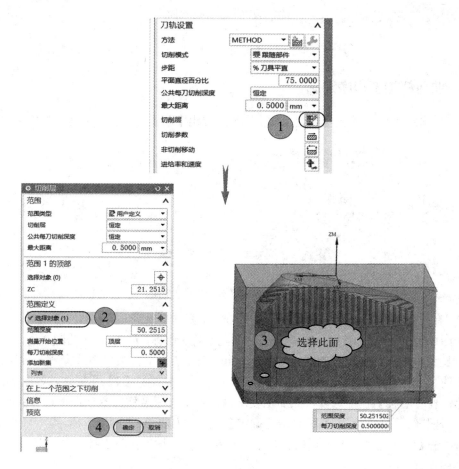

图 3-51　选择加工底面

(4) 其余参数都不需要修改,直接单击 ![]【生成】图标计算出新的加工路径。单击【确定】退出型腔铣,如图 3-52 所示。

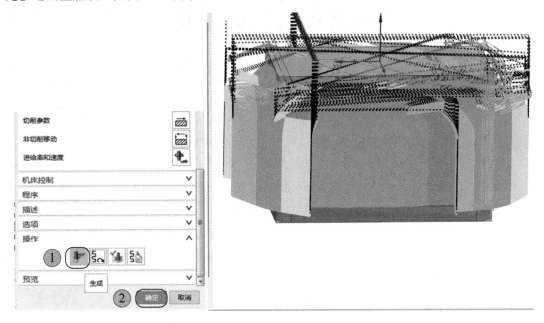

图 3-52　生成第二序粗加工程序

3.5.3　曲面精加工程序编制

(1) 单击图 3-53 所示的【创建工序】图标,弹出【创建工序】对话框,如图 3-54 所示。

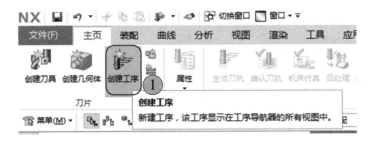

图 3-53　单击【创建工序】图标

(2) 在【类型】下拉菜单中选择【mill_contour】曲面铣选项,如图 3-54 圆圈 2 所示,出现如图 3-55 所示的对话框。选择 ![]【区域轮廓铣】图标,如图 3-55 圆圈 3 所示。在【程序】下拉菜单中选择刚建好的【上面】程序组,【刀具】下拉菜单中选择预先设置的【R3球刀】的合金球头铣刀,【几何体】下拉菜单中选择建立好的【2】几何体,如图 3-55 圆圈 4 所示。在【名称】栏中可以按照加工要求输入一个程序名称,本项目在这里就不做专门修改,按照默认名称填写,然后单击【确定】按钮,如图 3-55 圆圈 5 所示。

(3) 在【区域轮廓铣】对话框中,单击 ![]【指定切削区域】图标,窗选图中所示区域,如图 3-56 所示。

图 3-54　选择【mill_contour】曲面铣选项

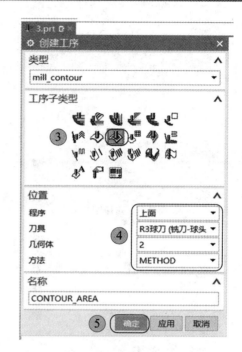

图 3-55　创建【区域轮廓铣】工序

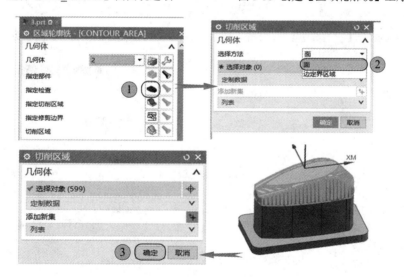

图 3-56　指定切削区域选项

(4) 完成切削区域设置后，选择【驱动方法】中【区域铣削】，如图 3-57 所示，再单击 ⚙【编辑】图标对曲面进行二次开粗参数设置，如图 3-58 所示。

(5) 检查【工具】中【刀具】以及【刀轴】设置，如错误，则需对其修改，如图 3-59 所示。刀轨设置中包含的【切削参数】和【非切削移动】无需修改。

(6) 单击 🔧【进给率和速度】图标，打开【进给率和速度】对话框，设置【主轴速度】为 "4000 rpm"，单击右侧 📋【计算器】图标。设置【进给率】为 "1500 mmpm"，【进刀】设置为 "70%"，然后单击【确定】按钮退出【进给率和速度】对话框，如图 3-60 所示。

图 3-57 【驱动方法】选项设置

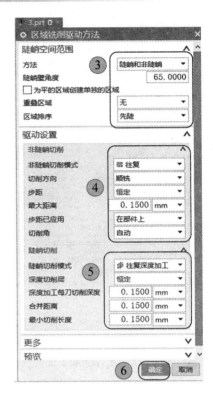

图 3-58 区域铣削参数设置

图 3-59 【刀具】、【刀轴】参数设置

图 3-60 【进给率和速度】参数设置

(7) 其余参数都不需要修改，直接单击 ⏩【生成】图标计算出新的加工路径。单击【确定】退出区域轮廓铣，如图 3-61 所示。

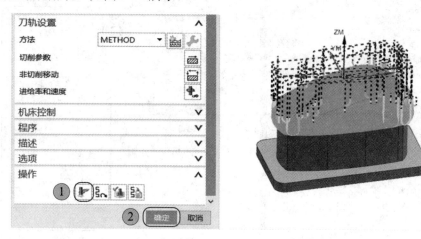

图 3-61 生成区域轮廓铣加工路径

3.5.4 曲面一次圆弧清根程序编制

(1) 复制上一步精铣曲面程序【区域轮廓铣】，内部粘贴到【2】几何体下，如图 3-62 所示。

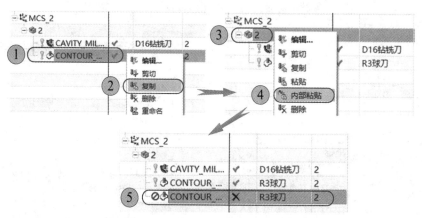

图 3-62 复制粘贴精加工区域轮廓铣程序

(2) 双击打开刚复制过来的 ⏩CONTOUR_AREA_COPY【区域轮廓铣】图标，打开【区域轮廓铣】对话框。单击【驱动方法】下的【方法】，更改为【清根】，再单击 🖉【编辑】图标，对清根参数进行设置并把【参考刀具】更改为上一次精加工【R3 球刀】，如图 3-63 所示。

(3) 单击【工具】中【刀具】进行刀具设置，将之前 R3 合金球头铣刀更改为提前创建好的【R2 球刀】，如图 3-64 所示。

(4) 单击 🐝【进给率和速度】图标，设置【主轴速度】为"6000 rpm"，单击右侧 🔢【计算器】图标。其他参数无需更改，然后单击【确定】按钮退出【进给率和速度】对话框，如图 3-65 所示。

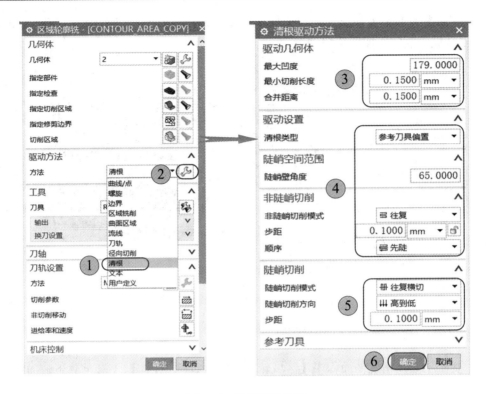

图 3-63　编辑清根设置

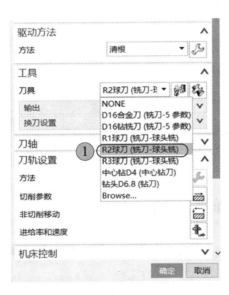

图 3-64　刀具的更改设置

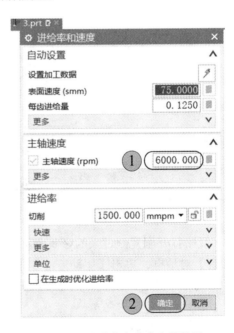

图 3-65　进给率和速度参数设置

(5) 单击 ▶【生成】图标计算新的加工路径。单击【确定】退出区域轮廓铣，如图 3-66 所示。

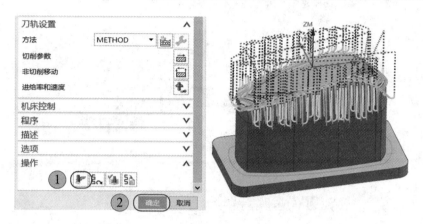

图 3-66 生成区域轮廓铣(清根)加工路径

3.5.5 曲面二次圆弧清根程序编制

同第一次清根加工方法相同，复制第一次清根加工程序，粘贴到【2】几何体下。单击 【编辑】图标对清根参数设置，将【参考刀具】改为【R2】球头铣刀，再单击 【进给率和速度】图标设置【主轴速度】为"8000 rpm"，单击右侧 【计算器】图标。其他参数无需更改。最后，单击 【生成】图标计算出新的加工路径。单击【确定】退出区域轮廓铣。生成的加工路径图如图 3-67 所示。

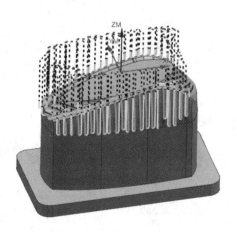

图 3-67 参考刀具设置和精加工清根路径

3.5.6 精加工侧面和底面程序编制

(1) 复制【底面】程序组中的【精铣壁】加工程序，内部粘贴到【2】几何体下，如图

3-68 所示。

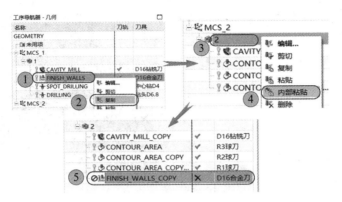

图 3-68　复制粘贴【精铣壁】加工程序

(2) 双击 FINISH_WALLS_COPY × D16合金刀 图标打开【精铣壁】对话框。单击几何体，更改为【2】。再单击 【指定部件边界】图标对加工边界进行更改，如图 3-69 所示。

图 3-69　几何部件边界设置

(3) 打开【部件边界】对话框，单击【添加新集】，选择【曲线】，如图 3-70 所示。

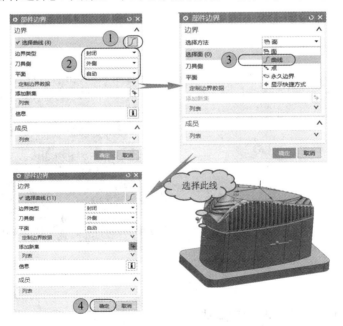

图 3-70　部件边界的选择

(4) 单击【指定底面】图标，弹出【平面】对话框，更改选择加工底面，如图 3-71 所示。

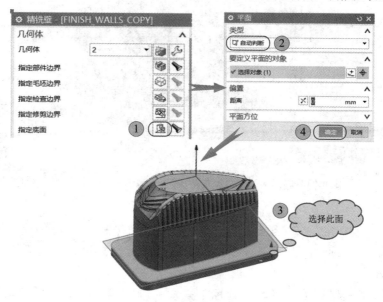

图 3-71 指定底面的选择

(5) 单击【切削层】图标弹出【切削层】对话框，【类型】选择【恒定】，【每刀切削深度】【公共】设置为"10 mm"，如图 3-72 所示。

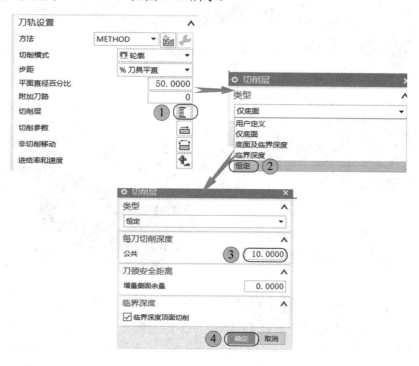

图 3-72 切削层更改参数设置

(6) 其余参数都不需要修改，直接单击 ▶【生成】图标计算出新的加工路径。单击【确定】退出区域轮廓铣，如图 3-73 所示。

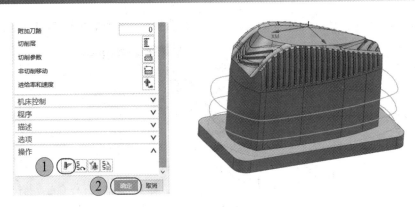

图 3-73　底面和侧壁精加工路径图　　　　　　　　项目三第二序机床仿真

3.6　生成 G 代码文件

（1）按住计算机键盘【Ctrl】键，鼠标左键选择【底面】程序组下所有程序，单击【后处理】图标，如图 3-74 所示，弹出【后处理】对话框。

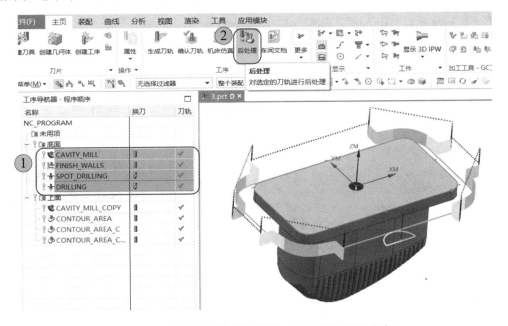

图 3-74　选择【底面】所有程序，单击【后处理】图标

（2）在【后处理器】选项中选择【FANUC0i】后处理文件，单击【输出文件】选项下的【浏览以查找输出文件】图标，弹出【指定 NC 输出】对话框(首先在 D 盘创建【nc】文件夹)，选择 D:\nc 目录，输入文件名为"1"，单击【OK】返回【后处理】对话框。确定文件名位置为 D:\nc\1，文件扩展名为 .nc。单击【确定】退出设置，如图 3-75 所示。

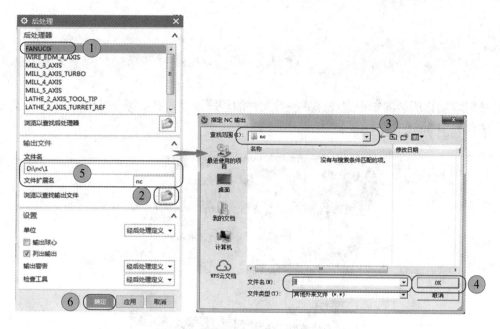

图 3-75　设置后处理文件位置及名称

(3) 单击【确定】后弹出【多重选择警告】对话框，单击【确定】按键将所有程序输出在一个程序组下显示，如图 3-76 所示。

(4) 弹出 G 代码文件并在 D 盘 nc 文件夹下生成 1.nc 文件，并且底面程序组下所有程序前面都显示出绿色对勾表示已经出完 G 代码文件，未出 G 代码文件加工程序前显示为黄色感叹号，如图 3-77 所示。

图 3-76　弹出【多重选择警告】对话框　　　　图 3-77　生成 G 代码文件

(5) 以相同的方法生成另一程序组的加工程序。

编程操作视频

项目三编程操作视频

课 后 练 习

按照本项目所学的知识完成课后练习文件的程序编制，课后练习见图 3-78 和光盘 "练习" 文件夹中 3.prt 文件。

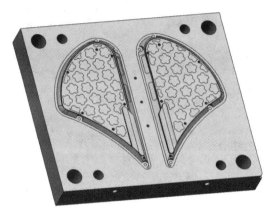

图 3-78　项目三课后练习图

项目四　自行车尾灯注塑模具定模型芯的程序编制

案例说明 📝

本项目以自行车尾灯注塑模具定模型芯为案例，讲解其加工工艺、加工方法的选择、切削刀具选择以及曲面相同多型腔的加工程序编制。

学习目标 📝

通过学习自行车尾灯注塑模具定模型芯的编程，读者应了解和掌握 NX 软件三轴同曲面多型腔零件的加工编程方法(曲面区域轮廓铣、曲面清根铣等)，以及曲面相同多型腔加工程序的复制和镜像程序的编制。做到举一反三，触类旁通。

学习任务 📝

曲面类零件是各类模具加工中较为常用的零件，加工形状较为复杂，且大多为单件加工，常应用于注塑模具、冲压模具、压铸模具、吹塑模具以及复杂的曲面零件加工等。通过本案例使读者掌握运用 NX 软件对相同多型腔曲面程序的复制和镜像编制加工的方法。

自行车尾灯注塑模具定模型芯三维图如图 4-1 所示。

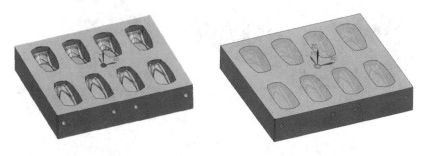

图 4-1　自行车尾灯注塑模具定模型芯三维图

4.1　自行车尾灯注塑模具定模型芯的加工工艺规程

加工工艺规程是描述每步加工过程的，一般包括被加工的区域、加工类型(平面铣、曲面铣、孔加工等)、工序内容描述、零件装夹、所需刀具及完成加工所必需的其他信息。

自行车尾灯注塑模具定模型芯最大外形尺寸为 246 mm × 222 mm × 40 mm，材料为模具钢 P20，此产品为多型腔单件加工。为便于铣削，在加工前对毛坯原料进行六面精加工铣磨，使其外形达到上述尺寸要求。

4.1.1 案例工艺分析

图 4-1 所示为自行车尾灯注塑模具定模型芯零件图，此工件只需加工一面型腔，无需对外轮廓进行加工和多次装夹即可完成工件的制作，为保证加工后零件的尺寸精度和形位公差，在编程前首先要确定好工件的加工工序方案。

其加工工序如下：

使用虎钳装夹加工零件长方底部。为了防止在加工过程中工件松动，夹持部位尽量大于工件总厚度的一半。安装毛坯后一定要测量工件的露出高度以确保安全生产。此次加工距离为 18.5 mm，加工时要注意加工部位的尺寸公差，按照图纸要求完成工件加工，如图 4-2 所示。此工件为相同多型腔零件，由于各个型腔相同，因此只需完成一个型腔的编程，然后通过程序的平移复制和程序的镜像，即可完成其他腔的编程。

项目四装夹方式

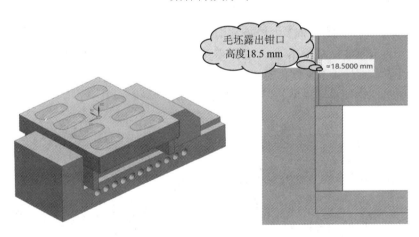

图 4-2　工序安装示意图

4.1.2 案例加工刀具的选择

本案例中自行车尾灯注塑模具定模型芯的最小加工圆角为 R1.006mm，如图 4-3 所示，故选用 R1 球头硬质合金铣刀进行加工。型芯内轮廓均由曲面构成，首先对其进行粗加工，由于粗加工时要去除大量的材料，加工时刀具承受的力会很大，所以要选取直径比较大的镶片钻铣刀 D16R0.4。这样加工时就不容易发生振刀、断刀、崩刃的现象。然后使用 D6合金铣刀进行二次开粗清理粗加工。由于型芯内轮廓型腔面均为曲面，因此再选用 R3 和

R2 球头硬质合金铣刀对其分别进行曲面精加工、加工后残余圆角部分余量清根加工。最后使用 R1 球头硬质合金铣刀精铣清根。

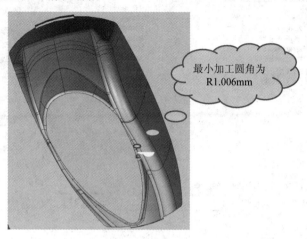

图 4-3　最小加工圆角示意图

　　按照上文分析的零件加工工艺方案和切削刀具的选择方式，合理安排零件的加工工艺过程。按照先粗后精、先面后孔、基准统一的原则设计本案例的加工工艺过程单，如表 4-1 所示。

表 4-1　自行车尾灯注塑模具定模型芯工艺过程单

工序号	顺序号 (刀具号)	加工机床	工序内容	工序名	刀具名称
1	1	加工中心	粗加工	型腔铣	D16R0.4
1	2	加工中心	二次开粗	型腔铣	D6
1	3	加工中心	型腔曲面精加工	型腔铣	R3
1	4	加工中心	清角加工	区域轮廓铣(清根)	R2
1	5	加工中心	清角加工	区域轮廓铣(清根)	R1

4.2　打开模型文件进入加工模块

　　打开随书配套光盘，在"例题"文件夹中打开模型文件，并且进入加工模块。

　　(1) 启动 NX 12.0，单击左上角 📂【打开】按钮，在"查找范围"选项中选择光盘"例题"文件夹中 4.prt 文件，如图 4-4 所示。

自行车尾灯定模型芯的程序编制

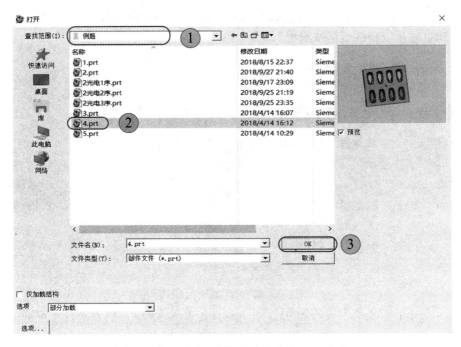

图 4-4　打开光盘"例题"文件夹中 4.prt 文件

(2) 单击【应用模块】按钮,再单击【加工】按钮(也可以直接单击键盘快捷键 Ctrl+Alt+M)启动 UG NX12.0 加工模块, 如图 4-5 所示。

(3) 默认选择【CAM 会话配置】中的【cam_general】, 在【要创建的 CAM 组装】对话框中选择【mill_contour】曲面铣, 单击【确定】按钮, 如图 4-6 所示, 进入曲面铣加工界面。如图 4-6 所示。

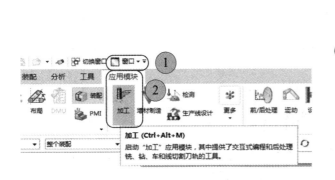

图 4-5　进入 UG NX 12.0 加工模块　　　　图 4-6　【加工环境】对话框设置

4.3　建立父节点组

父节点组包括几何视图、机床视图、程序顺序视图和加工方法视图。

(1) 几何视图。几何视图可定义"加工坐标系"方向和安全平面，并设置"部件""毛坯"和"检查"几何体等参数。

(2) 机床视图。机床视图可定义切削刀具，并指定铣刀、钻头、和车刀等，并保存与刀具相关的数据，以用作相应后处理命令的默认值。

(3) 程序顺序视图。程序顺序视图能够把编好的程序按组排列在文件夹中，并按照从上到下的先后顺序排列加工程序。

(4) 加工方法视图。加工方法视图用来定义切削方法类型(粗加工、精加工、半精加工)。如"内公差""外公差"和"部件余量"等参数。

4.3.1　创建加工坐标系

在【几何】视图菜单中创建加工坐标系的操作步骤如下：

(1) 将【工序导航器】切换到【几何】视图页面，如图 4-7 所示。

(2) 右击【MCS_MILL】图标，如图 4-7 圆圈 3 所示，选择【重命名】选项，然后设置工序坐标系名称为【MCS_1】，双击打开【MCS 铣削】对话框，如图 4-8 所示。

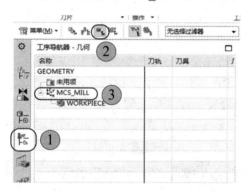

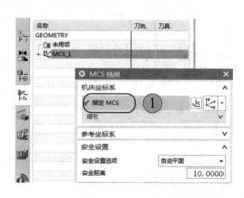

图 4-7　【几何】视图切换　　　　　　　图 4-8　【MCS 铣削】对话框

(3) 由于加工原料为六面精磨料，为保证各面加工留量均匀，所以把坐标系放在毛坯中间位置。单击 【坐标系对话框】图标，如图 4-9 圆圈 1 所示。在弹出的【坐标系】对话框中选择【对象的坐标系】选项，如图 4-9 圆圈 2 所示，其功能是自动设置坐标为所选择平面的中心点位置。单击工件上表面方框自动捕捉工件的上表面中心点坐标位置，如图 4-9 圆圈 3 所示。在抬刀安全平面没有干涉物的情况下可以选择默认状态【自动平面】，【安全距离】为"10 mm"，如图 4-9 圆圈 4 所示。如有干涉物则可把安全距离设置为"50 mm～100 mm"。最后单击【确定】按钮退出【MCS 铣削】对话框，如图 4-9 圆圈 5 所示。

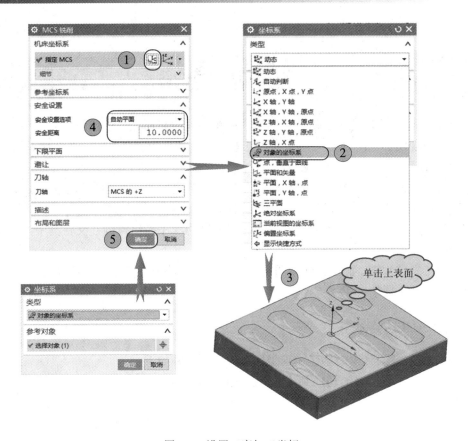

图 4-9　设置工序加工坐标

4.3.2　创建部件几何体

在【几何】视图菜单中创建加工部件几何体、零件毛坯及检查几何体的操作步骤如下：

(1) 右击 WORKPIECE 图标选择【重命名】，设置工序工件名称为"1"，双击打开【工件】对话框，如图 4-10 所示。

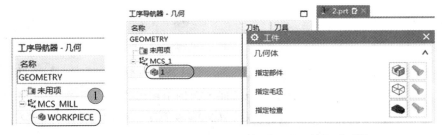

图 4-10　选择【WORKPIECE】图标弹出【工件】对话框

(2) 单击 【选择或编辑部件几何体】图标，如图 4-11 圆圈 1 所示。弹出【部件几何体】对话框后单击被加工工件，使其成为橘黄色，然后单击【确定】按钮退出【部件几何体】对话框，如图 4-11 圆圈 2 所示。

(3) 单击 【选择或编辑毛坯几何体】图标，如图 4-11 方框 1 所示。弹出【毛坯几何体】对话框，在【类型】选项中选择【包容块】选项，如图 4-12 方框 2 所示。设置毛坯尺寸单边增加"0 mm"，如图 4-12 方框 3 所示。单击【确定】按钮，如图 4-12 方框 4 所示。返回【工件】对话框。然后再单击【确定】按钮退出【工件】对话框完成设置，如图 4-12 方框 5 所示。

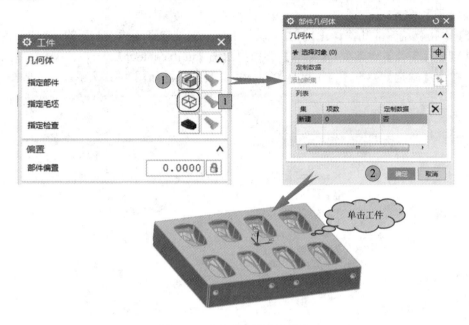

图 4-11　建立【部件几何体】

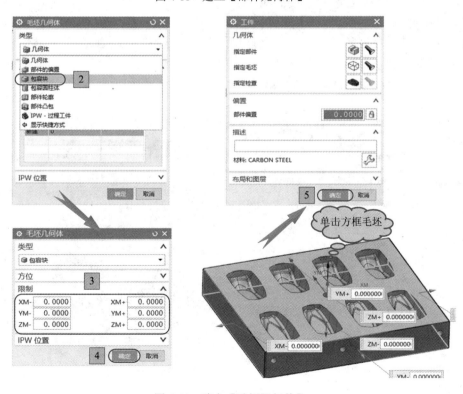

图 4-12　建立【毛坯几何体】

(4) 在选择完毛坯后,毛坯方框在编辑加工程序中便再无用处,我们可以把它隐藏起来,避免后序操作中产生误操作问题。其具体方法为左键单击毛坯方框,按键盘快捷键 Ctrl+B。如果想返回显示,则按键盘快捷键 Ctrl+Shift+K,然后左键选择要恢复的图形退出【显示】设置。

(5) 单击已编辑的几何体【1】图标,单击鼠标右键选择【插入】下的【创建几何体】,弹出【创建几何体】对话框,【类型】选择曲面铣【mill_contour】,再单击 【铣削区域】图标,单击【确定】弹出【铣削区域】对话框。设置方法如图 4-13 所示。

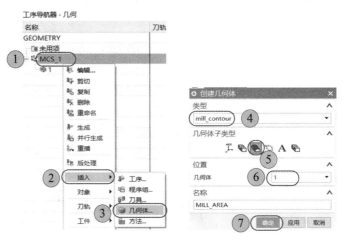

图 4-13　创建几何体【1】下的【铣削区域】

(6) 单击 【指定切削区域】图标,打开【切削区域】选择对话框,单击【选择方法】【面】,然后单击选择对象,用鼠标窗选加工区域,单击【确定】完成设定,其他设置无需更改,如图 4-14 所示。

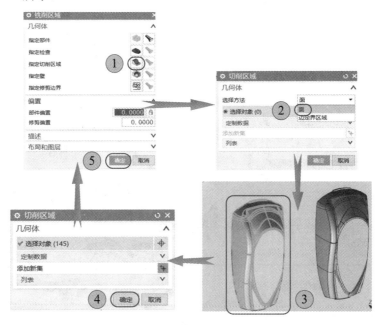

图 4-14　创建【切削区域】

(7) 创建完成后的切削区域【MILL_AREA】如图 4-15 所示。

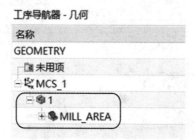

图 4-15　创建后的切削区域【MILL AREA】在【几何】视图下的位置

4.3.3　创建刀具

在【机床】视图下创建加工刀具步骤如下：

(1) 单击 【机床】视图图标，将【工序导航器】切换到【机床】视图页面，如图 4-16 所示。

(2) 单击 【创建刀具】图标，如图 4-17 所示，弹出【创建刀具】对话框。

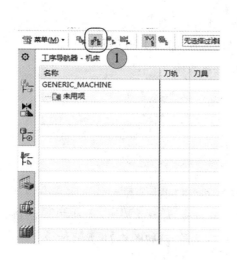

图 4-16　【机床】视图

图 4-17　【创建刀具】对话框

(3) 选择 【平底刀】图标，创建平底刀。【名称】位置输入"D16R0.4"(代表直径为 16 mm 圆角、半径为 0.4 mm 的镶片钻铣刀)，如图 4-18 所示。

(4) 【直径】设置为"16"，【下半径】设置为"0.4"，【刀具号】【补偿寄存器】【刀具补偿寄存器】三项均设置为"1"(此数值代表刀具、刀具半径补偿和刀具长度补偿号，为避免发生撞机问题，最好设置为相同数字)。单击【确定】完成刀具建立，如图 4-19 所示。

其他刀具在编辑加工程序前，按照给定参数自行设置。

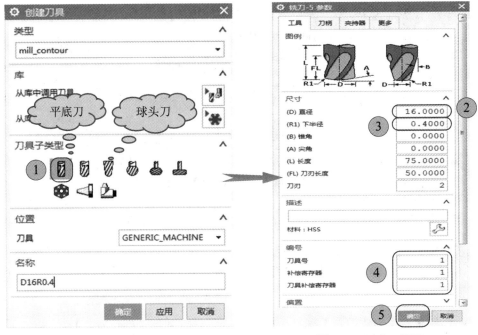

图 4-18　创建平底刀　　　　　　　　图 4-19　平底刀设置

4.3.4　创建程序组

在【程序顺序】视图中创建加工程序组文件夹，操作步骤如下：

(1) 将【工序导航器】切换到【程序顺序】视图页面，如图 4-20 所示。

图 4-20　【工序导航器】切换到【程序顺序】视图页面

(2) 双击【PROGRAM】程序图组文字，修改文件名为"底面"(或使用右键单击【PROGRAM】程序组，选择【重命名】也可实现更改名称)，如图 4-21 所示。

图 4-21　双击修改程序组名称

(3) 保存文件。

4.4 自行车尾灯注塑模具定模型芯的粗加工

由前面的加工分析得出自行车尾灯注塑模具定模型芯加工特点为相同多型腔，加工面只有一面，为此零件底部一个型腔。无需为每一型腔都进行程序编写，只需通过程序的平移复制和程序的镜像，即可完成其他腔的编程。以下为此工件粗加工的程序编制方法。

4.4.1 粗加工程序编制

NX 软件在建立有模型图和毛坯的基础上编程时，使用最为简单有效的开粗程序就是曲面加工里的型腔铣，使用型腔铣可以完成绝大多数零件的开粗工作。由于此原料为六面精磨料，所以在粗加工时工件底面的表面是不需要加工的，只需要加工其凹下去的型腔即可。为了更好地体现出加工程序的先后顺序，本案例在编程时全部使用【程序顺序】视图来完成程序的编制。

图 4-22 单击【创建工序】图标

(1) 单击【创建工序】图标，如图 4-22 所示，弹出【创建工序】对话框。

(2) 在【类型】下拉菜单中选择【mill_contour】曲面铣选项，如图 4-23 圆圈 2 所示。选择 【型腔铣】图标，在【程序】下拉菜单中选择刚建好的【底面】程序组，【刀具】下拉菜单中选择【D16R0.4】的镶片钻铣刀，【几何体】下拉菜单中选择建立好的【MILL_AREA】切削区域，在【名称】栏中可以按照加工要求输入一个程序名称，本案例在这里就不做专门修改，按照默认名称填写，然后单击【确定】按钮，如图 4-24 所示。

图 4-23 选择【mill_contour】曲面铣选项

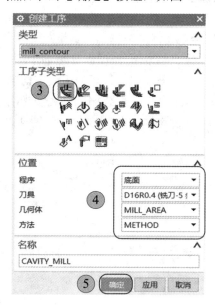

图 4-24 创建【型腔铣】工序

(3) 只要在【几何】视图正确设定【MILL_AREA】切削区域，那么在弹出的【型腔铣】对话框中，【指定部件】、【指定毛坯】、【指定切削区域】选项应显示为灰色，右侧🖉【显示】图标为彩色，单击它可显示已选择的几何体和切削区域部分。如果进入型腔铣后还能选择部件、毛坯和切削区域，则说明【几何】视图中的【MILL_AREA】没有设定，或进入程序前没有选择几何体为【MILL_AREA】。本次加工中【指定检查】和【指定修剪边界】不需要设置，如图 4-25 所示。

(4) 打开【工具】菜单下拉箭头，显示已选择的加工刀具为"D16R0.4"镶片钻铣刀。(注：此项工作前序选择正确的情况下可忽略)。

(5) 打开【刀轴】菜单下拉箭头，显示默认刀轴为【+ZM 轴】，三轴加工中心刀轴一般使用+ZM 轴，只有在使用多轴机床加工时才会修改此项(注：三轴加工编程时不用选择此选项，使用默认设置即可)，如图 4-26 所示。

图 4-25　几何体设置对话框　　　　　　　　图 4-26　工具和刀轴选项

(6)【刀轨设置】为型腔铣参数设置的主要内容。【切削模式】选项中一般常用🔲跟随部件和🔄跟随周边 两种方式，【跟随部件】适合加工开放轮廓的工件，可以使刀具从外向内加工并从工件外下刀。【跟随周边】更适合加工封闭轮廓的工件，可以使刀具从内向外加工，减少型腔加工时的下刀位置变化。本案例为封闭轮廓的加工，所以在【切削模式】中选择【跟随周边】的加工方法，如图 4-27 圆圈 1 所示。

(7) 粗加工 XY 方向，刀具步距一般使用刀具直径的 70%～80%；精加工时，步距使用刀具直径的 50%以下。80%～100%的 XY 方向步距一般情况下不推荐使用。首先，刀具步距太大的话，每次切削时相当于满刀切削，刀具受力过大会影响刀具寿命和机床精度。其次，如果步距太大，那么加工完底面光洁度很低，会出现接刀痕。因此在【平面直径百分比】选项中设置数值为"75%"，如图 4-28 圆圈 2 所示。

图 4-27　选择【跟随周边】　　　　　　　　图 4-28　XYZ 方向步距量设置

(8) D16R0.4 钻铣刀粗加工时的 Z 方向步距一般情况下取值 0.3 mm～0.7 mm 每层。【最大距离】选项设置的就是刀具 Z 方向的每层步距量，因此设置中间数值"0.5 mm"每层，如图 4-28 圆圈 3 所示。由于设置了区域铣削，故 ☰【切削层】不需要设置。

(9) 单击 ☲【切削参数】图标，如图 4-28 圆圈 4 所示，打开【切削参数】对话框。

(10) 在【策略】标签中设置【切削顺序】为【深度优先】，【深度优先】会按照不同区域分别由上往下加工，从而可以缩短抬刀和过刀路径，减少加工时间。【层优先】会按照同一深度在不同区域跳刀加工，从而会增加很多抬刀路径，增加加工时间。一般情况下优先选择【深度优先】，如图 4-29 所示。

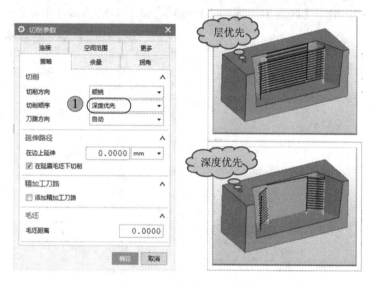

图 4-29　【策略】标签设置

(11) 选择【余量】标签，设置【部件侧面余量】参数为"0.2 mm"(注：粗加工刀具余量一般设置为 0.2mm)，单击【确定】按钮退出【切削参数】对话框，如图 4-30 所示。

图 4-30　【余量】标签设置

(12) 单击 ⊞【非切削移动】图标，如图 4-31 圆圈 1 所示，打开【非切削移动】对话框。

(13) 设置进刀参数首先要设置【封闭区域】【进刀类型】，选择【螺旋】。在型腔尺寸能够满足螺旋下刀时，一般选用螺旋下刀的方式进刀。在型腔尺寸不能满足螺旋下刀时，一般选用斜线下刀或直线下刀的方式进行。螺旋下刀直径选用刀具【直径】的"50%"，【斜坡角度】设置为"5 度"，【高度】设置为"1 mm"，【最小斜坡长度】设置为"0"(注：加工刀具能够直线下刀时可以输入数字为"0"。如果加工刀具不能直线下刀，则此数字最小不能小于 50，否则会发生撞机事故)，如图 4-32 圆圈 2 所示。【开放区域】设置进刀【长度】为刀具的"50%"，抬刀【高度】设置为"1 mm"，如图 4-32 圆圈 3 所示。

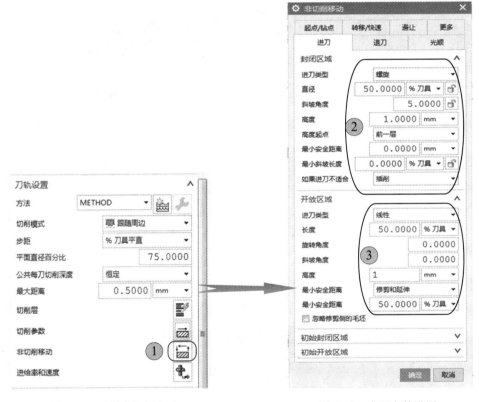

图 4-31　选择非切削移动　　　　　　图 4-32　进刀参数设置

(14) 单击【非切削移动】对话框【转移/快速】标签设置快速抬刀高度。为提高加工速度减少抬刀高度，把【区域之间】和【区域内】的【转移类型】都改为【前一平面】，并且把抬刀【安全距离】都设置为"1 mm"。但大部分快速移刀都是在工件零平面以下进行，所以要求机床的 G00 运动必须是两点间的直线运动，不能是两点间的折线运动，否则会发生撞机事故。加工前一定要在 MDI 下输入"G00 走斜线观察机床"的移动方式，如果不对，则需要修改机床参数。或者按照 NX12.0【转移/快速】的初始设置方式，把【转移类型】全部设置为【安全距离-刀轴】。每次提刀都回到安全平面会增加加工时间。最后单击【确定】按钮退出【非切削移动】对话框，具体参数设置如图 4-33 所示。

图 4-33　【转移/快速】标签设置

(15) 单击 <!-- icon --> 【进给率和速度】图标，打开【进给率和速度】对话框。设置【主轴速度】为"2500 rpm"(注：输入"2500"后一定要按后面的【计算器】图标，否则会报警)。【进给率】【切削】输入"1500 mmpm"，【进刀】输入"70%切削"。单击【确定】退出【进给率和速度】对话框，如图 4-34 所示。

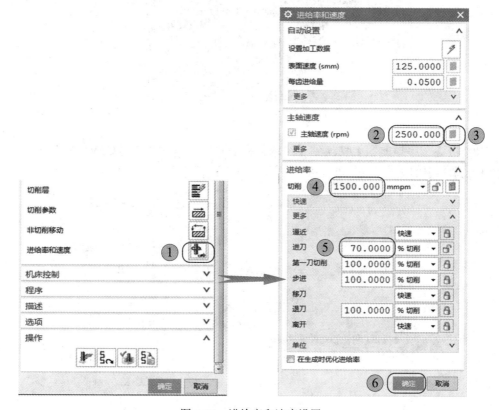

图 4-34　进给率和速度设置

(16) 单击 ▶ 【生成】图标计算加工路径。单击【确定】退出型腔铣，如图 4-35 所示。

图 4-35　生成刀具路径

(17) 沿 X 轴正方向测量两个型腔之间的投影距离为 57 mm，即每个型腔沿 X 轴的间距均为此值，如图 4-36 所示。

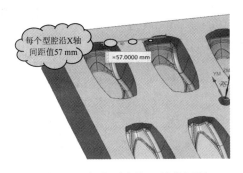

图 4-36　每个型腔沿 X 轴的间距

(18) 将【工序导航器】切换到【机床】视图页面，右键单击【CAVITY_MILL】图标选择【对象】中的【变换】，如图 4-37 所示，单击打开【变换】对话框。

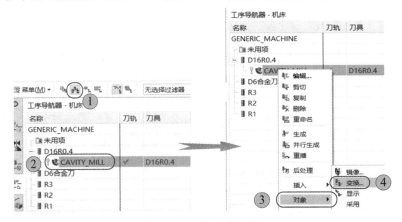

图 4-37　变换导航设置

(19)【类型】选择【平移】，如图 4-38 圆圈 1 所示。【变换参数】选择【增量】，【XC 增量】值输入"57"，【结果】选择【复制】，【非关联副本数】输入"3"(注：因为图纸中沿 XC 方向还有 3 个型腔没有编程，所以输入值为 3)。单击【确定】，完成复制，复制程序结果如图 4-39 所示。

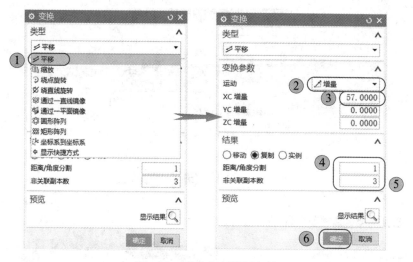

图 4-38　变换复制设置参数

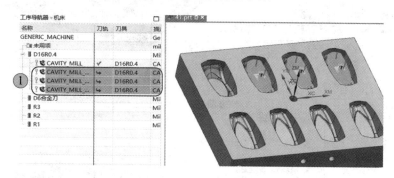

图 4-39　复制程序结果

(20) 按住【Shift】，用鼠标选中已经编好的 4 腔程序，如图 4-40 圆圈 1 所示。右键选择【对象】中的【变换】，单击打开【变换】对话框。

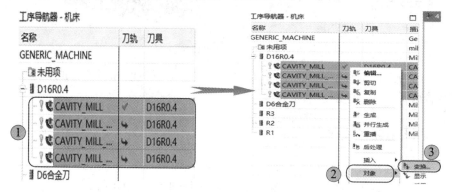

图 4-40　变换导航设置

(21) 在【变换】对话框中【类型】选项中选择【通过—平面镜像】,【变换参数】中单击【平面选择】图标,弹出【平面】对话框,【类型】选择【XC-ZC 平面】,其他参数不变,单击【确定】完成平面选择。回到【变换】对话框,【结果】选择【复制】,单击【确定】,完成镜像,如图 4-41 所示。程序镜像结果如图 4-42 所示。

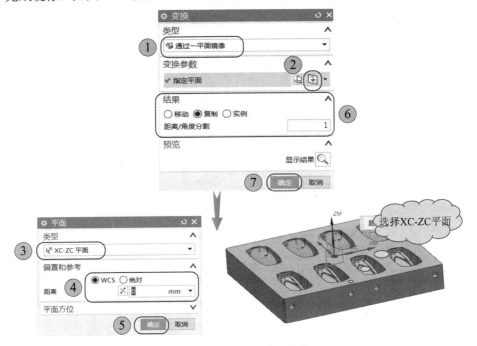

图 4-41　变换镜像设置参数

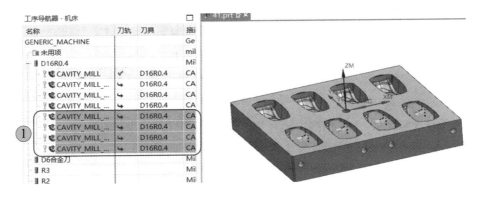

图 4-42　程序镜像结果

以上程序为工件所有型腔第一次开粗程序,主要通过复制和镜像完成。这样有效地节约了程序的编写时间,提高了编程效率。

4.4.2　二次开粗程序编制

由于第一次开粗后,型腔余量较大,尤其是一些圆角位置。为保证精铣顺利进行,因此用先前创建的 D6 硬质合金铣刀在第一次开粗的基础上,进行二次粗加工。

(1) 在【机床】视图中右键单击刚编好的【型腔铣】程序，选择【复制】选项，如图 4-43 所示。然后右键单击【D6 合金刀】选择【内部粘贴】，如图 4-44 所示。形成一个新的提示错误的【型腔铣】程序，如图 4-45 所示。

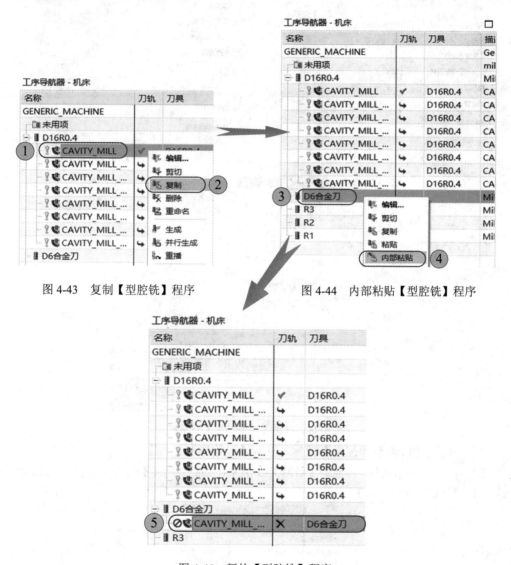

图 4-43　复制【型腔铣】程序　　　　图 4-44　内部粘贴【型腔铣】程序

图 4-45　新的【型腔铣】程序

(2) 双击打开新复制过来的【型腔铣】程序。设置 Z 向分层【最大距离】为 "0.2 mm"，如图 4-46 所示。

(3) 单击 ⏢【切削参数】图标，修改【切削参数】设置，如图 4-47 所示。

(4) 在【切削参数】对话框中【空间范围】标签中设置工件剩余毛坯。在【过程工件】下拉菜单中选择【使用基于层的】选项，设置本次加工毛坯为上次加工剩余的部分，在【剩余铣】中此选项默认为【使用基于层的】选项，【型腔铣】和【剩余铣】的不同只有这一个选项。然后单击【确定】按钮，如图 4-48 所示。

图 4-46 设置 Z 向分层为"0.2 mm"每刀

图 4-47 单击【切削参数】图标

图 4-48 设置【切削参数】对话框

(5) 单击 【进给率和速度】图标，打开【进给率和速度】对话框，设置【主轴速度】为"3500 rpm"，单击转速右侧 【计算器】图标，最后单击【确定】按钮退出【进给率和速度】对话框，如图 4-49 所示。

图 4-49 设置主轴转速

(6) 单击 【生成】图标计算加工程序，生成二次开粗加工路径，单击【确定】按钮生成刀具路径，如图 4-50 所示。

图 4-50　生成刀具路径

(7) 其余型腔的二次开粗程序同样采用平移复制和镜像复制来完成，可参考 4.4.1(粗加工程序编制)中步骤(17)～步骤(21)。完成结果如图 4-51 所示。

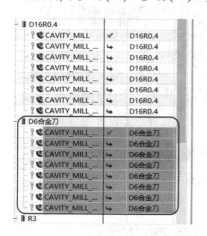

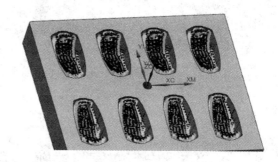

图 4-51　二次开粗复制和镜像生成程序

4.4.3　三次开粗程序编制

(1) 在【机床】视图中右键单击刚编好的二次开粗 D6 合金刀【型腔铣】程序，选择【复制】选项，然后右键单击之前创建好的【R6 合金刀】，选择【内部粘贴】，形成一个新的提示错误的【型腔铣】程序，结果如图 4-52 所示。

(2) 双击打开新复制过来的【型腔铣】程序。更改【刀轨设置】中的【步距】为【恒定】，【最大距离】为 "0.3 mm"，其他参数设置不变，如图 4-53 所示。

(3) 单击 【生成】图标计算加工程序，生成三次开粗加工路径，单击【确定】按钮生成刀具路径，如图 4-54 所示。

(4) 其余型腔的三次开粗程序，同样采用平移复制和镜像复制来完成。

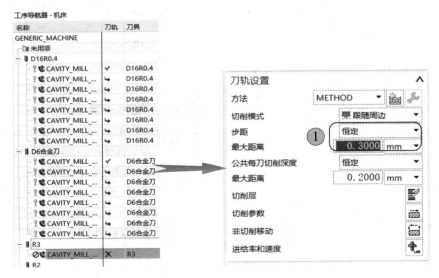

图 4-52　三次开粗复制程序　　　　　　图 4-53　更改步距设置

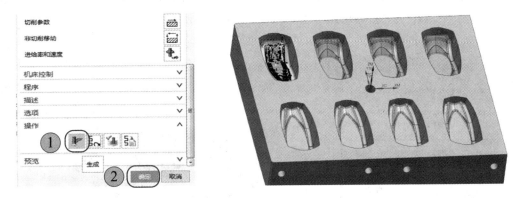

图 4-54　三次开粗刀具路径

4.5　自行车尾灯注塑模具定模型芯的精加工

4.5.1　曲面精加工程序编制

(1) 单击【创建工序】图标，弹出【创建工序】对话框，如图 4-55 所示。

图 4-55　单击【创建工序】图标

(2) 在【类型】下拉菜单中选择【mill_contour】曲面铣选项，如图 4-56 圆圈 2 所示。选择 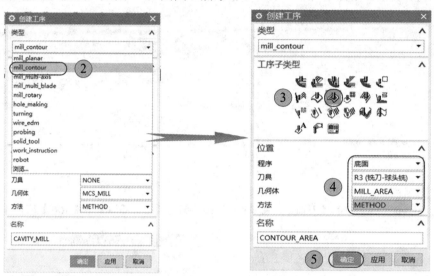 【区域轮廓铣】图标，如图 4-57 圆圈 3 所示。在【程序】下拉菜单中选择刚建好的【底面】程序组，【刀具】下拉菜单中选择预先设置的【R3(铣刀—球头铣)】，【几何体】下拉菜单中选择建立好的【MILL_ARER】切削区域，如图 4-57 圆圈 4 所示。在【名称】栏中可以按照加工要求输入一个程序名称，在这里就不做专门修改，按照默认名称填写，然后单击【确定】按钮，如图 4-57 圆圈 5 所示。

图 4-56　选择【mill_contour】曲面铣选项　　　　图 4-57　创建【区域轮廓铣】工序

(3) 进入【区域轮廓铣】对话框，选择【驱动方法】【方法】的【区域铣削】，如图 4-58，再单击 【编辑】图标对曲面进行精加工参数设置，如图 4-59 所示。

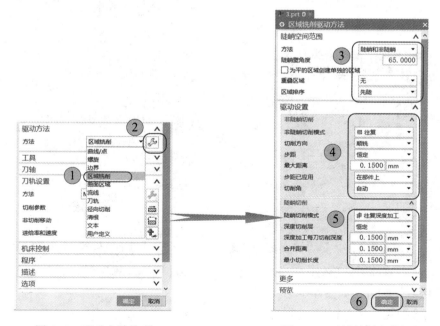

图 4-58　驱动方法选项　　　　　　　　图 4-59　区域铣削参数设置

(4) 检查【工具】中【刀具】以及【刀轴】设置，如图 4-60 所示。

(5) 单击【刀轨】设置中包含的【切削参数】图标，弹出【切削参数】对话框，分别对【策略】和【余量】进行修改，如图 4-61 所示和图 4-62 所示，设置完成后单击【确定】。

(6) 单击 ⊕【进给率和速度】图标，设置【主轴速度】为 "6000 rpm"，单击右侧 ▤【计算器】图标。设置【进给率】【切削】为 "1500 mmpm"，【进刀】设置为 "70%切削"，然后单击【确定】按钮退出【进给率和速度】对话框，如图 4-63 所示。

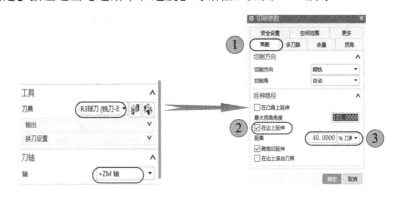

图 4-60 【刀具】、【刀轴】设置参数　　图 4-61　【切削参数】中【策略】的设置

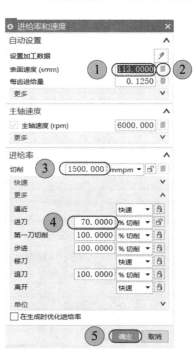

图 4-62　【切削参数】中【余量】标签的设置　　图 4-63　【进给率和速度】参数设置

(7) 其余参数都不需要修改，直接单击 ⊫【生成】图标计算出新的加工路径。单击【确定】退出区域轮廓铣，如图 4-64 所示。

(8) 其余型腔精加工程序，同样采用平移复制和镜像复制来完成。

图 4-64 生成区域轮廓铣加工路径

4.5.2 曲面精加工第一次清根程序编制

(1) 在【机床】视图下复制上一步【R3】球头铣刀精铣曲面程序【区域轮廓铣】，内部粘贴到【R2】球头铣刀下，如图 4-65 所示。

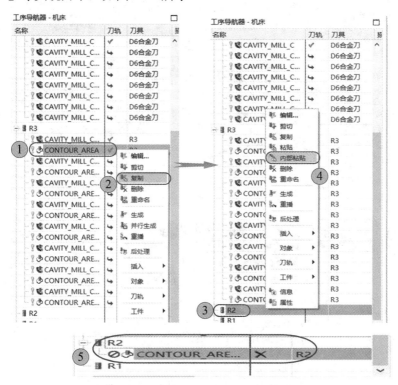

图 4-65 复制粘贴精加工【区域轮廓铣】程序

(2) 双击打开刚复制过来的【区域轮廓铣】程序，打开【区域轮廓铣】对话框。单击【驱动方法】下的【方法】，更改为【清根】，再单击 【编辑】图标，打开【清根驱动方法】对话框，对清根参数进行设置，如图 4-66 所示。

(3) 单击 【进给率和速度】图标，打开【进给率和速度】对话框，设置【主轴速度】为"8000 rpm"，单击右侧 【计算器】图标。其他参数无需更改，然后单击【确定】按钮退出【进给率和速度】对话框，如图 4-67 所示。

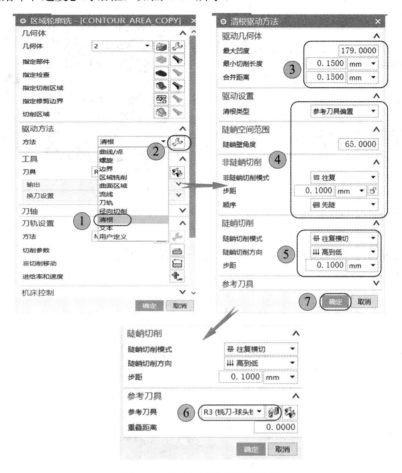

图 4-66　驱动方法清根参数设置

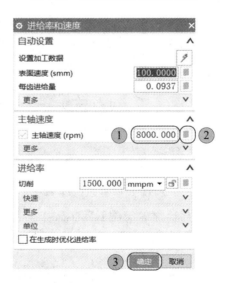

图 4-67　【进给率和速度】参数设置

（4）单击 【生成】图标计算出新的加工路径。单击【确定】退出【区域轮廓铣】。如图 4-68 所示。

图 4-68　生成区域轮廓铣(清根)加工路径

（5）其余型腔精加工清根程序，同样采用平移复制和镜像复制来完成，结果如图 4-69 所示。

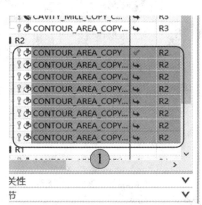

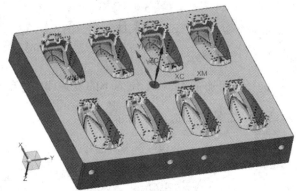

图 4-69　复制和镜像生成的清根刀路

4.5.3　曲面精加工第二次清根程序编制

（1）同第一次清根加工方法相同，在【机床】视图下复制上一步【R2】球头铣刀精铣曲面程序【区域轮廓铣】，内部粘贴到【R1】球头铣刀下。双击打开刚复制过来【区域轮廓铣】程序，单击 【编辑】图标对清根参数进行设置。将【参考刀具】改为【R2(铣刀—球头铣)】，如图 4-70 所示。再单击 【进给率和速度】图标，打开【进给率和速度】对话框，设置【主轴速度】为"10000 rpm"，单击右侧 【计算器】图标。其他参数无需更改。最后，单击 【生成】图标计算出新的加工路径。单击【确定】退出【区域轮廓铣】。

（2）其余型腔精加工清根程序，同样采用平移复制和镜像复制来完成。生成的加工路径如图 4-71 所示。

图 4-70　修改编辑里的参考刀具

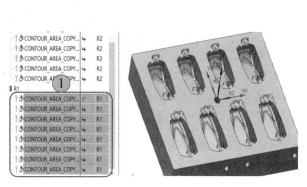

图 4-71　生成区域轮廓铣(清根)加工路径

4.6　生成 G 代码文件

(1) 选择开粗加工【D16】铣刀下的所有程序，单击 【后处理】图标，弹出【后处理】对话框，如图 4-72 圆圈 2 所示。

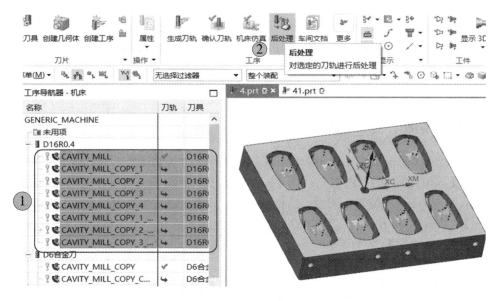

图 4-72　选择程序单击【后处理】图标

(2) 在【后处理器】选项中选择【FANUC0i】后处理文件，单击【输出文件】选项下的 📁【浏览以查找输出文件】图标，弹出【指定 NC 输出】对话框(首先在 D 盘创建 "nc" 文件夹)，选择 D:\nc 目录，输入文件名为 "1"，单击【OK】返回【后处理】对话框。确定文件名位置为 D:\nc\1，文件扩展名为 .nc。单击【确定】退出设置，如图 4-73 所示。

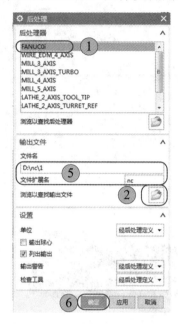
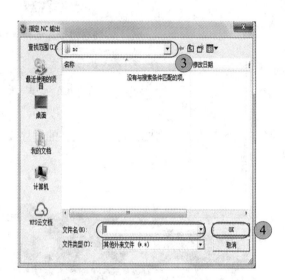

图 4-73 设置后处理文件位置及名称

(3) 单击【确定】后弹出【多重选择警告】对话框，单击【确定】按键将所有程序输出在一个程序组下显示，如图 4-74 所示。

(4) 弹出 G 代码文件并在 D 盘 nc 文件夹下生成 1.nc 文件，并且 D16R0.4 钻铣刀下所有程序前面都显示出绿色对勾，这表示已经出完 G 代码文件，未出的 G 代码文件加工程序前显示为黄色感叹号，如图 4-75 所示。

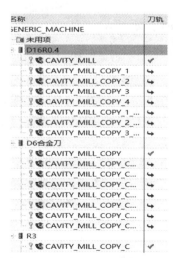

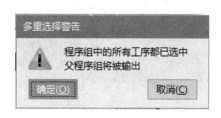

图 4-74 弹出【多重选择警告】对话框　　　图 4-75 生成 G 代码文件

(5) 以相同的方法生成其他程序组的加工程序。

编程操作视频

项目四编程操作视频

课 后 练 习

按照本课所学的知识完成课后练习文件的程序编制,课后练习见图 4-76 和光盘"练习"文件夹中 4.prt 文件。

图 4-76   项目四课后练习图

项目五　吹塑模具瓶体的程序编制

案例说明 ✍

本项目以吹塑模具瓶体编程加工为案例，讲解吹塑模具瓶体的加工工艺、加工方法的选择、刀具路径转换、大型模具切削刀具选择以及曲面加工编程的注意事项。

学习目标 ✍

通过学习吹塑模具瓶体模型的加工编程，读者应了解和掌握 NX 软件三轴曲面零件的加工编程方法。做到举一反三，触类旁通。

学习任务 ✍

曲面轮廓类零件是机械加工中常用的零件，加工造型比较多，且大多单件加工，常应用于汽车、医疗、玩具、模具等工业领域中。

本案例以吹塑模具瓶体加工编程为案例讲解 NX 软件三轴曲面编程。在这个过程中，重点是让读者学会并理解 NX 软件各种曲面加工程序的使用方法和应用，可以按照不同的方法和工艺生成加工程序。

5.1　吹塑模具瓶体的加工工艺规程

加工工艺规程是用来描述每步加工过程的，一般包括被加工的区域、加工类型(平面铣、曲面铣、孔加工等)、工序内容描述、零件装夹、所需刀具及完成加工所必需的其他信息。

吹塑模具瓶体外形尺寸为 530 mm × 302.8 mm × 120 mm，材料为 45#钢。在加工模具类零件前需要把锻料经过普铣和平磨加工至外形尺寸达到图纸要求，并且六面成 90 度直角关系。模具类工件大部分为单件加工，而且钢材外形铣削比较耗时，并且切削力较大和长期刷料对数控机床的精度也会有一定影响，所以模具类零件毛坯数控加工前都需要经过普通机床加工外形的工序，数控机床只加工模具型腔造型部分。

5.1.1　案例工艺分析

图 5-1 所示为吹塑模具瓶体模型，此工件需要加工正面型腔部分和左右两侧定位边。由于工件外形尺寸较大，虎钳的行程范围不能装夹住此工件，因此只能选用压板装夹。在

压板装夹的过程中需要调换压板位置，所以在此工件加工中，需要使用两次装夹来完成工件的加工。

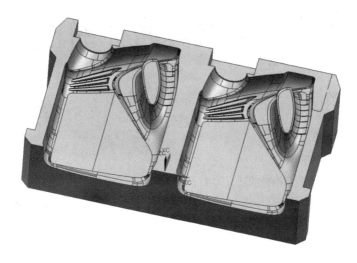

图 5-1　吹塑模具瓶体模型

为保证加工后零件的尺寸精度和形位公差，在编程前首先要确定好工件的加工工序方案。

工序一：

使用四个压板压住毛坯中间的部位，留出左右两侧加工部分，并且多留出加工时压板和刀柄之间的安全距离，以避免发生碰撞事故。安装位置如图 5-2 所示。注意：由于工件较大，开粗余量较多，压板装夹时必须使用四个压板装夹，否则可能会出现工件被铣动的情况。装夹后第一序使用铣刀，先将工件左右两侧面加工出来。

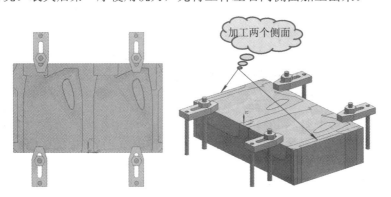

图 5-2　加工工序一装夹方式

项目五第一序装夹方式

工序二：

加工完第一序后先不要急于拆掉压板。先准备 4 个新的压板，按照如图 5-3 所示的方式压住工件四个凸起的部分，然后再拆掉工序一装夹时使用的四个压板。这样装夹的好处是在调换压板时，工件还被工序一的四个压板固定着，这时新装的压板就能够起到固定的作用，再拆掉工序一的压板就不会发生位移的问题了。

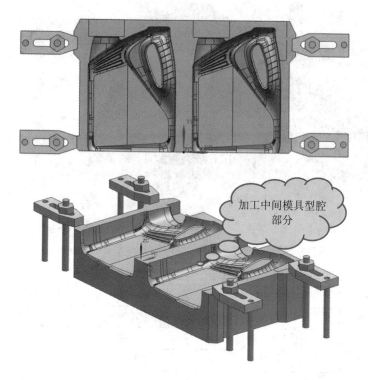

图 5-3　加工工序二装夹方式

项目五第二序装夹方式

5.1.2　案例加工刀具的选择

　　本例是吹塑模具瓶体模型的加工，最小凹圆角为半径为 1 mm，如图 5-4 所示。由于型腔深度较高，R1 球刀不能够加工到型腔底面，所以只能选用电火花机床完成最后的清角工作。在这里我们只加工到 R3 球刀即可。加工模型左右两侧面时，由于侧壁加工深度较大而且是直面，如果选用直柄刀加工，则当刀具加工到较深的位置时，铣削过程中会发生刀柄夹铁削的问题，从而造成刀具损坏，因此我们在加工时应选用刀头切削部分直径为 35mm，其余刀柄部分直径为 32mm 的铣刀去加工，就能尽量避免夹削问题的产生。加工模具型腔部分时，由于型腔内轮廓均是由曲面构成的，因此第一步要对其进行粗加工。由于工件尺寸较大、开粗余量较多，粗加工时为了更快地去除加工余量，所以要选取直径比较大的镶片面铣刀 D50R0.8 加工。第二步使用 D12R0.8 的镶片钻铣刀对其进行二次开粗，清理粗加工后残余的圆角部分余量。第三步使用 R5 球刀对其手柄部分和瓶体上部进行第三次开粗，清理上一把刀具加工后残余的部分。第四步使用 R5 球刀精加工型腔曲面轮廓。最后一步使用 R3 球刀清根加工图中细小的圆角部分，完成工件的制作。

　　按照上文分析的零件加工工艺方案和切削刀具的选择方式，合理安排零件的加工工艺过程。按照先粗后精、先面后孔、基准统一的原则设计本案例的加工工艺过程单，如表 5-1 所示。

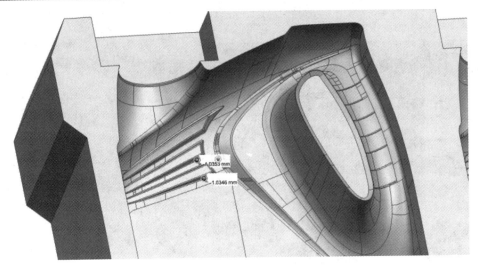

图 5-4　最小凹圆角示意图

表 5-1　吹塑模具瓶体工艺过程单

工序号	顺序号	加工机床	工序内容	工序名	刀具名称
1	1	加工中心	粗加工	深度轮廓铣	D35R0.8
1	2	加工中心	精加工侧壁	精铣壁	D25
2	3	加工中心	粗加工	型腔铣	D50R0.8
2	4	加工中心	二次开粗	剩余铣	D12R0.8
2	5	加工中心	二次开粗	剩余铣	R5
2	6	加工中心	精加工曲面	区域轮廓铣	R5
2	7	加工中心	清根铣	清根铣	R3
3	8	加工中心	右型腔加工	平移 2 组所有程序	同上

5.2　打开模型文件进入加工模块

打开配套光盘，在"例题"文件夹中打开模型文件，并进入加工模块。

吹塑模具瓶体零件的程序编制

（1）启动 NX 12.0，单击左上角 🖮【打开】按钮，在"查找范围"选项中选择光盘"例题"文件夹中 5.prt 文件，如图 5-5 所示。

（2）单击【应用模块】按钮，再单击【加工】按钮(也可以直接单击快捷键 Ctrl + Alt + M)进入 NX12.0 加工模块，如图 5-6 所示。

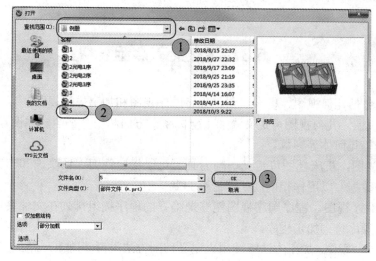

图 5-5　打开光盘"例题"文件夹中 5.prt 文件

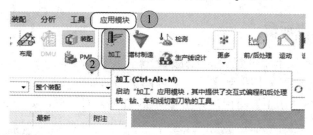

图 5-6　进入 NX 12.0 加工模块

（3）默认选择【CAM 会话配置】【cam_general】，在【要创建的 CAM 组装】对话框中选择曲面铣【mill_contour】按钮，单击【确定】按钮进入【平面铣加工】界面，如图 5-7 所示。

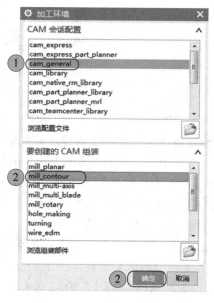

图 5-7　【加工环境】对话框设置

5.3 建立父节点组

父节点组包括几何视图、机床视图、程序顺序视图和加工方法视图。

(1) 几何视图。几何视图可定义"加工坐标系"方向和安全平面，并设置"部件""毛坯"和"检查"几何体等参数。

(2) 机床视图。机床视图可定义切削刀具。可以指定铣刀、钻头、和车刀等，并保存与刀具相关的数据，以用作相应后处理命令的默认值。

(3) 程序顺序视图。程序顺序视图能够把编好的程序按组排列在文件夹中，并按照从上到下的先后顺序排列加工程序。

(4) 加工方法视图。加工方法视图用来定义切削方法类型(粗加工、精加工、半精加工)。例如，"内公差""外公差"和"部件余量"等参数在此设置。

5.3.1 创建加工坐标系

在【几何】视图菜单中创建加工坐标系的操作步骤如下：

(1) 将【工序导航器】切换到【几何】视图页面，如图 5-8 所示。

(2) 双击【MCS_MILL】图标，弹出【MCS 铣削】对话框，如图 5-9 所示。

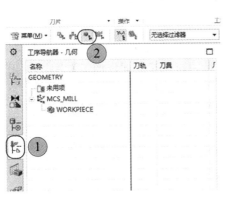

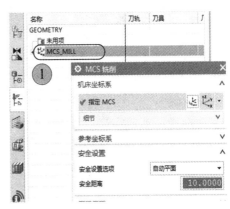

图 5-8　【几何】视图切换　　　　　图 5-9　【MCS 铣削】对话框

(3) 由于毛坯材料前期经过普通机床机加工处理，表面形状和外形尺寸都符合图纸要求，为便于合模，所以把加工坐标系放在毛坯中间位置。单击 【坐标系对话框】图标，如图 5-10 圆圈 1 所示。选择【对象的坐标系】选项，如图 5-11 圆圈 2 所示。【对象的坐标系】选项功能是，自动设置坐标为所选择平面的中心点位置。单击毛坯表面，自动捕捉出工件的上表面中心点坐标位置，如图 5-12 圆圈 3 所示。此工件为压板装夹，在刀具左右移动的过程中，如果抬刀安全高度过低则可能会产生撞刀现象。因此在【安全设置】选项中设置安全平面为【自动平面】,【安全距离】设置为"150 mm"，超过压板高度。最后单击【确定】按钮退出【MCS 铣削】对话框，如图 5-10 圆圈 4 和圆圈 5 所示。

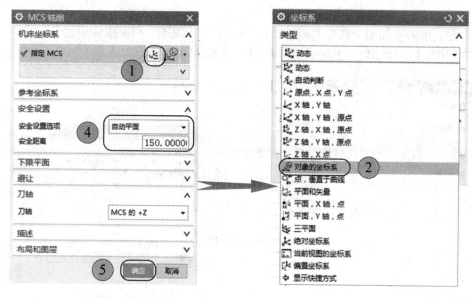

图 5-10　选择【坐标系对话框】图标　　　　图 5-11　选择【对象的坐标系】选项

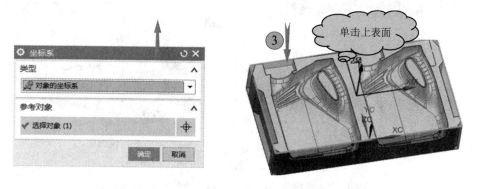

图 5-12　单击工件上表面方框设置坐标系

5.3.2　创建部件几何体

在【几何】视图菜单中创建加工部件几何体、零件毛坯、检查几何体的操作步骤如下：

(1) 双击 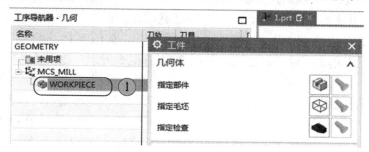 WORKPIECE 图标，打开【工件】对话框，如图 5-13 所示。

图 5-13　选择【WORKPIECE】图标弹出【工件】对话框

(2) 单击 【选择或编辑部件几何体】图标，如图 5-14 圆圈 1 所示。弹出【部件几何体】对话框，单击被加工工件使其成橘黄色，如图 5-14 圆圈 2 和圆圈 3 所示，然后单击【确定】按钮退出【部件几何体】对话框，如图 5-14 圆圈 4 所示。

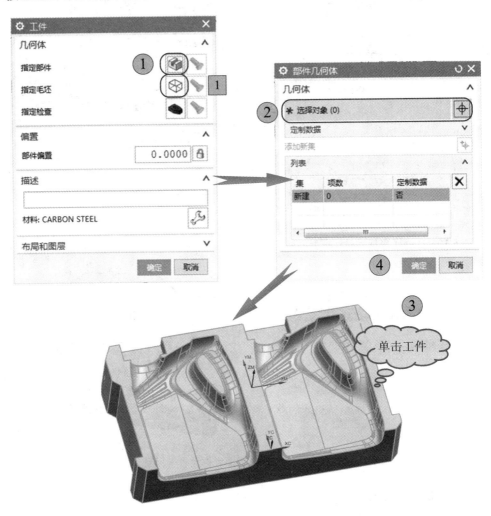

图 5-14　建立部件几何体

(3) 单击 【选择或编辑毛坯几何体】图标，如图 5-14 方框 1 所示。弹出【毛坯几何体】对话框，在【类型】选项中选择【几何体】选项，如图 5-15 方框 2 所示。单击图中给定的毛坯，如图 5-15 方框 3 所示。单击【确定】按钮，如图 5-15 方框 4 所示。返回【工件】对话框然后再单击【确定】按钮，如图 5-15 方框 5 所示。退出【工件】对话框完成设置。

(4) 在选择完毛坯后，毛坯方框在编辑加工程序中便再无用处，可以把它隐藏起来，避免后序操作中产生误操作问题。其具体方法为：左键单击毛坯方框，按键盘快捷键 Ctrl＋B。如果想返回显示，则按键盘快捷键 Ctrl＋Shift＋K，然后左键选择要恢复的图形退出【显示】设置。

图 5-15　建立【毛坯几何体】

5.3.3　创建刀具

在【机床】视图下创建粗加工刀具 D35R0.8 钻铣刀的操作步骤如下:

(1) 单击 [icon]【机床】视图图标,将【工序导航器】切换到【机床】视图页面,如图 5-16 圆圈 1 所示。

(2) 单击 [icon]【创建刀具】图标,如图 5-17 圆圈 2 所示,弹出【创建刀具】对话框。

(3) 选择 [icon]【平底刀】图标,创建平底刀如图 5-17 圆圈 3 所示。【名称】位置输入 "D35R0.8"(代表直径为 35 mm 圆角半径为 0.8 mm 的镶片钻铣刀)。

(4) 【直径】设置为 "35",【下半径】设置为 "0.8",【刀具号】【补偿寄存器】【刀具补偿寄存器】三项均设置为 "1"(此数值代表刀具、刀具半径补偿和刀具长度补偿号,为避免发生撞机问题,最好设置为相同数字)。单击【确定】完成刀具建立,如图 5-18 所示。单击【刀柄】标签设置【刀柄直径】为 "32 mm"、【刀柄长度】为 "150 mm"、【锥柄长度】为 "0",如图 5-19 所示。

其他刀具在编辑加工程序前,按照给定的参数自行设置。

图 5-16 【机床】视图

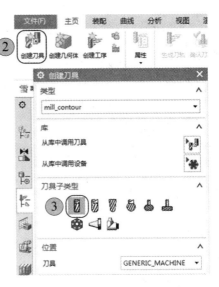

图 5-17 【创建刀具】对话框

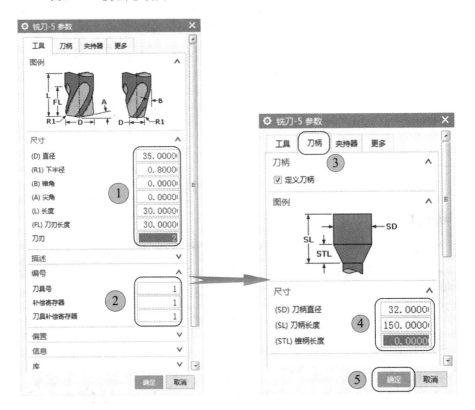

图 5-18 创建 φ35 钻铣刀

图 5-19 创建刀柄为"32 mm"

5.3.4 创建程序组

在【程序顺序】视图中创建加工程序组文件夹，操作步骤如下：

(1) 将【工序导航器】切换到【程序顺序】视图页面，如图 5-20 所示。

(2) 双击【PROGRAM】程序组文件夹，如图 5-20 圆圈 2 所示。将程序组名称修改为"1"(或使用右键单击【PROGRAM】程序组，选择【重命名】也可实现更改名称)，如图5-21 圆圈 3 所示。

图 5-20 【工序导航器】切换到【程序顺序】视图页面   图 5-21 双击修改程序组名称

(3) 保存文件。

5.4 吹塑模具瓶体第一序加工

吹塑模具瓶体第一序加工部分为模型左右两侧面。以下为第一序加工的程序编制方法。

5.4.1 粗加工程序编制

加工模型左右两侧面余量最大的地方宽度为 15 mm，被加工面宽度小于粗加工刀具直径。对于这种开放轮廓，且加工余量小于刀具直径的工件，最好的加工方法为深度轮廓铣。深度轮廓铣可以实现开放轮廓的往复加工，减少不必要的抬刀路径，从而提高加工效率。

为了更好地体现出加工程序的先后顺序，接下来的编程工作全部在【程序顺序】视图中来完成。

(1) 单击【创建工序】图标，如图 5-22 所示，弹出【创建工序】对话框。

图 5-22 单击【创建工序】图标

(2) 在【类型】下拉菜单中选择【mill_contour】曲面铣选项，如图 5-23 所示。选择【深度轮廓铣】图标，如图 5-24 圆圈 2 所示，在【程序】下拉菜单中选择刚建好的【1】程序组，【刀具】下拉菜单中选择【D35R0.8】的镶片钻铣刀，【几何体】下拉菜单中选择建立好的【WORKPIECE】几何体，在【名称】栏中可以按照加工要求输入一个程序名称，本案例在这里不做专门修改，按照默认名称填写，然后单击【确定】按钮，如图 5-24 圆圈3 和圆圈 4 所示。

图 5-23　选择【mill_contour】曲面铣选项

图 5-24　创建【型腔铣】工序

(3) 在弹出的【深度轮廓铣】对话框中，左键单击【指定切削区域】图标，如图 5-25 所示，选择模型左右两侧需要加工的实体表面。

(4) 打开【工具】菜单中【刀具】下拉箭头，显示出已选择的加工刀具为【D35R0.8(铣刀-5)】(注：此项工作前序选择正确的情况下可忽略)。

(5) 打开【刀轴】下拉菜单，显示，默认刀轴为【+ZM】轴，如图 5-26 圆圈 4 所示，三轴加工中心刀轴一般使用+ZM轴，只有在使用多轴机床加工时才会修改此项(注：此选项三轴加工编程时使用默认设置即可)，如图 5-26 圆圈 3 和圆圈 4 所示。

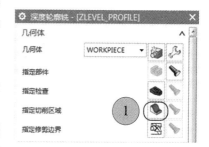

图 5-25　几何体设置对话框

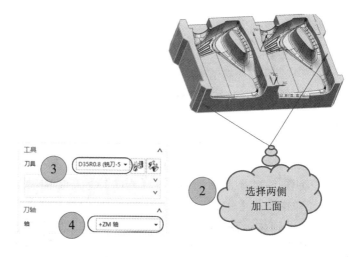

图 5-26　【工具】和【刀轴】选项

(6) 在【刀轨设置】中，修改 Z 轴分层【最大距离】参数为 "0.5mm" 每层，其他参数保持不变，如图 5-27 圆圈 1 所示。

(7) 左键单击 【切削层】图标，如图 5-27 圆圈 2 所示，设置加工【范围深度】为 "119.8mm"，使刀具加工到最后一刀时和工作台留有 0.2 mm 余量，这样可以避免铣削到机床工作台表面的加工事故。单击【确定】退出【切削层】设置，如图 5-28 圆圈 3 所示。

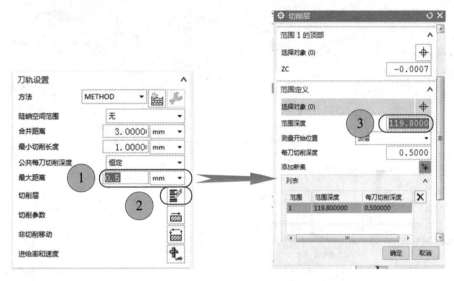

图 5-27　Z 方向分层 "0.5 mm" 每层　　　　图 5-28　设置加工【范围深度】

(8) 单击【切削参数】图标打开【切削参数】对话框，如图 5-29 圆圈 1 所示。在【策略】标签中设置【切削方向】为【混合】，【切削顺序】为【始终深度优先】，如图 5-29 圆圈 2 所示。设置这两个选项可以使走刀路径变为反复的形式，并且能减少不必要的抬刀。然后在【余量】标签中设置【部件侧面余量】为 "0.2mm"，【内公差】与【外公差】都为 "0.01 mm"，如图 5-29 圆圈 3 和圆圈 4 所示。单击【确定】退出【切削参数】设置，如图 5-29 圆圈 5 所示。

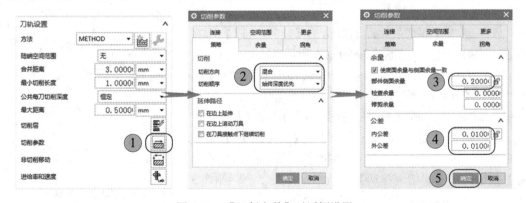

图 5-29　【切削参数】对话框设置

(9) 单击 【非切削移动】图标，如图 5-30 所示，打开【非切削移动】对话框。

(10) 设置进刀参数，首先要设置【封闭区域】的下刀参数，型腔内下刀一般采用螺旋方式下刀。螺旋下刀的加工原理：在型腔尺寸能够满足螺旋下刀时，首先选用螺旋进刀的方式下刀。在型腔尺寸不能够满足螺旋下刀要求时，再选用斜线下刀或直线下刀的方式。

螺旋下刀【直径】选用刀具直径的"50%"，【斜坡角度】设置为"5 度"，【高度】设置为"1 mm"，【最小斜坡长度】设置为"0"(注：加工刀具能够直线下刀时最小斜坡长度可以输入数字"0"。如果加工刀具不能够直线下刀，则此数字最小不能小于 50，否则会发生撞机事故)，如图 5-31 圆圈 2 所示。【开放区域】设置进刀【长度】为刀具的"50%"，抬刀【高度】设置为"1mm"，以减少抬刀距离。具体参数设置如图 5-31 圆圈 3 所示。

图 5-30　选择【非切削移动】　　　　图 5-31　进刀参数设置

(11) 单击【非切削移动】对话框【转移/快速】标签设置快速抬刀高度，如图 5-32 圆圈 1 所示。为提高加工速度减少抬刀高度，把【区域内】的【转移类型】改为【前一平面】，【安全距离】设置为"1"。由于第一序加工时工件中间装有压板，所以不能使用前一平面的设置方法。在【区域之间】的【转移类型】必须设置为【安全距离-刀轴】，如图 5-32 圆圈 2 所示，否则会发生撞机事故。最后单击【确定】按钮退出【非切削移动】对话框，如图 5-32 圆圈 3 所示。

(12) 单击  【进给率和速度】图标，打开【进给率和速度】对话框。设置【主轴速度】为"2000 rpm"(注：输入"2000"后一定要按后面的【计算器】图标，否则会报警)。输入【进给率】【切削】为"1500 mmpm"，输入

图 5-32　【转移/快速】标签设置

【更多】里【进刀】为"70%切削"，然后单击【确定】退出【进给率和速度】对话框，如图5-33所示。

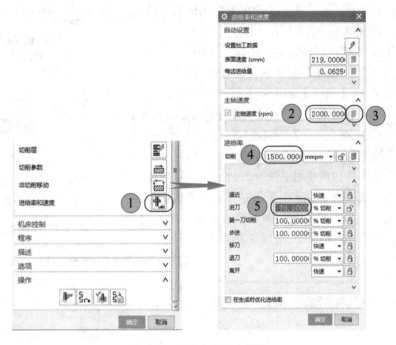

图5-33　进给率和速度设置

(13) 单击 ┣┿【生成】图标计算加工路径。单击【确定】退出【深度轮廓铣】，如图5-34所示。

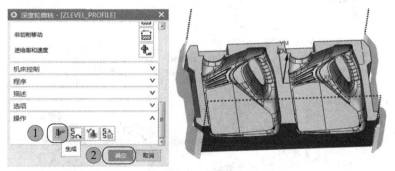

图5-34　生成刀具路径

5.4.2　精加工侧面程序编制

使用φ25镶嵌合金刀精加工零件左右两侧外轮廓的编制方法如下：

(1) 创建一把直径为25 mm、长150 mm、刀具名称为D25合金、刀具号为2号的镶嵌合金平底刀。使用φ25合金刀精加工外形轮廓。单击 【创建工序】图标，打开【创建工序】对话框，在【类型】下拉菜单中选择【mill_planar】平面铣选项，如图5-35圆圈1所示。【工序子类型】选项中选择 【精铣壁】的加工方法，如图5-35圆圈2所示。【程

序】选项中选择【1】程序组,【刀具】选项中选择【D25
合金】铣刀,【几何体】选项中选择【WORKPIECE】,如
图 5-35 圆圈 3 所示。最后单击【确定】按钮进入【精铣
壁】对话框,如图 5-35 圆圈 4 所示。

(2) 单击🔲【指定部件边界】图标,如图 5-36 圆圈
1 所示,打开【部件边界】对话框。在【边界】【选择方
法】选项中,选择【曲线】的方式创建加工边界,如图
5-36 圆圈 2 所示。设置【边界类型】为【开放】,【刀具侧】
为【左】,【平面】为【指定】,如图 5-36 圆圈 3 和圆圈 4
所示。单击工件上表面为加工起始平面,如图 5-36 圆圈 5
所示,然后手动选择轮廓曲线,选择开放曲线要从顺铣方
向的起点位置选择曲线,并且在选择完一根边界线后单击
鼠标滚轮中键,然后再选择下一条边界线。单击【确定】
退出设置,如图 5-36 圆圈 6、圆圈 7 和圆圈 8 所示。

图 5-35 创建精铣壁程序

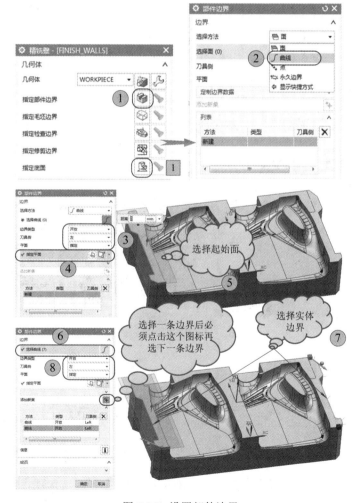

图 5-36 设置部件边界

(3) 单击 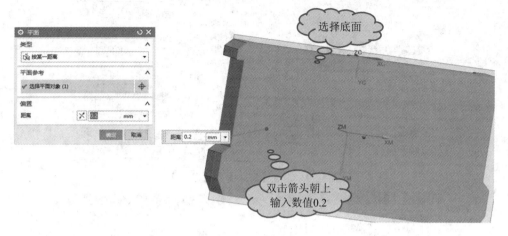 【指定底面】图标，如图 5-36 方块 1 所示，弹出【平面】对话框。设置加工底面，选择模型底面并且双击蓝色箭头使之朝上，输入 "0.2 mm"，将刀具最后一刀抬起 0.2 mm 以避免铣削机床工作台，如图 5-37 所示。

图 5-37 选择加工底面

(4) 单击 【切削层】图标，弹出【切削层】对话框，在【类型】下拉菜单中选择【恒定】的切削方式，每层的切削高度是恒定的，【每刀切削深度】【公共】设置为 "20mm"，单击【确定】按钮退出设置，如图 5-38 所示。

图 5-38 设置切削层

(5) 单击 【切削参数】图标，在弹出的【切削参数】对话框的【余量】标签中，设置【部件余量】和【最终底面余量】全部为 "0"，设置【内公差】和【外公差】数值全部为 "0.01"，单击【确定】按钮退出设置，操作步骤如图 5-39 所示。

(6) 单击 【非切削移动】图标，如图 5-39 方块 1 所示，打开【非切削移动】对话框，设置【进刀类型】为【线性】，并且打开半径补偿功能，具体设置方法如图 5-40 所示。

(7) 单击 【进给率和速度】图标，如图 5-39 三角 1 所示，打开【进给率和速度】对话框，设置【主轴速度】为 "2000 rpm"，【进给率】【切削】为 "500 mmpm"（精加工侧壁时走刀速度不宜过快，应控制在 "500 mmpm" 以下，否则加工表面粗糙度过大会影响加工质量）。然后单击【确定】按钮退出【进给率和速度】对话框，如图 5-41 所示。

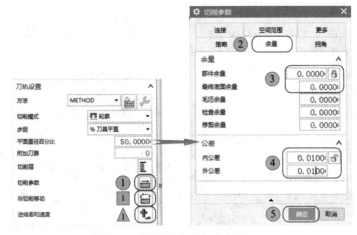

图 5-39　设置【余量】和【公差】

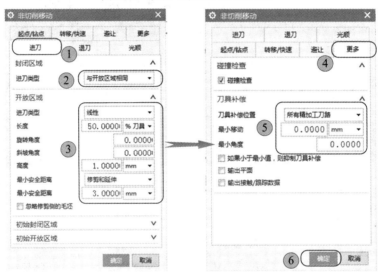

图 5-40　设置非切削移动对话框

图 5-41　设置主轴转速和进给速度

（8）单击 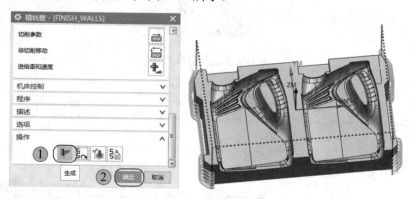【生成】图标计算加工程序，生成【精铣壁】精加工两侧壁路径，单击【确定】按钮退出【精铣壁】设置，如图 5-42 所示。

图 5-42　生成【精铣壁】加工程序

（9）仿真编辑完的加工程序，单击左上方 【程序顺序】视图图标，在【程序顺序】视图界面全选所有编完的加工程序，如图 5-43 所示。

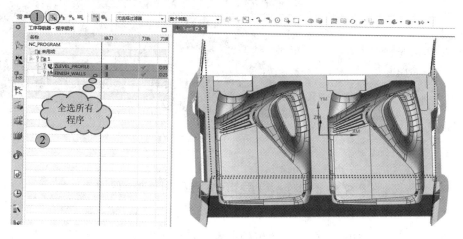

图 5-43　【程序顺序】视图全选所有加工程序

（10）单击【主页】 【确认刀轨】图标，打开【刀轨可视化】对话框，如图 5-44 所示。

图 5-44　单击【确认刀轨】图标

（11）单击【3D 动态】，切换【模拟动画】为【三维立体模型】，然后单击下方 ▶【播放】图标，完成工件底面路径的加工仿真，如图 5-45 所示。

项目五第一序机床仿真

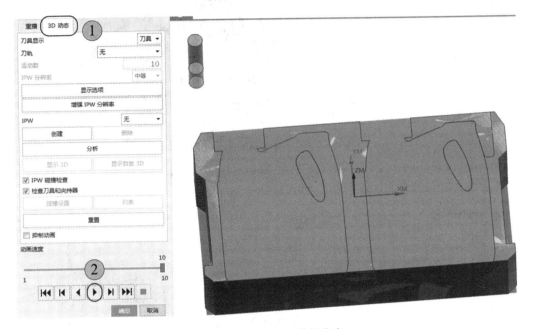

图 5-45　工序一加工路径仿真

5.5　吹塑模具瓶体第二序加工

加工完第一序后，按照前面工艺分析的要求，将压板改为左右两侧装夹留出型腔中间部分，进行第二序铣削加工。在编辑第二序加工程序前，先在【程序顺序】视图下创建一个新的程序组命名为【2】。

5.5.1　粗加工程序编制

首先使用型腔铣的加工方法完成零件的粗加工程序编制。为了更好地体现出加工程序的先后顺序，接下来在编程时全部使用【程序顺序】视图完成程序的编制。

(1) 创建直径为 50 mm、圆角半径为 R0.8、刀长为 100 mm 的面铣刀一把，输入名称"D50R0.8"，刀具号为 3 号。

(2) 创建【型腔铣】程序，在【类型】下拉菜单中选择【mill_contour】曲面铣选项，如图 5-46 所示。单击 ⭕【型腔铣】图标，在【程序】下拉菜单中选择新建的【1】程序组，【刀具】下拉菜单中选择【D50R0.8】的面铣刀，【几何体】下拉菜单中选择建立好的【WORKPIECE】几何体，然后单击【确定】按钮，如图 5-47 所示。

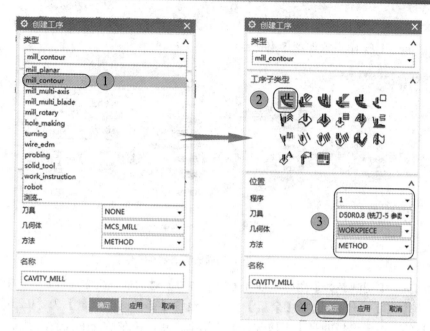

图 5-46 选择【mill_contour】曲面铣选项   图 5-47 创建【型腔铣】工序

（3）在弹出的【型腔铣】对话框中左键单击 【指定切削区域】图标，选择模具单一型腔曲面。由于模型较大而且曲面比较多，如果将两个型腔曲面全部选中加工，那么计算程序的工作量非常大，计算速度也会非常慢。为了提高程序的计算速度，本例只编辑一个型腔程序，然后采用平移程序的方法制作另一型腔的程序，如图 5-48 所示。

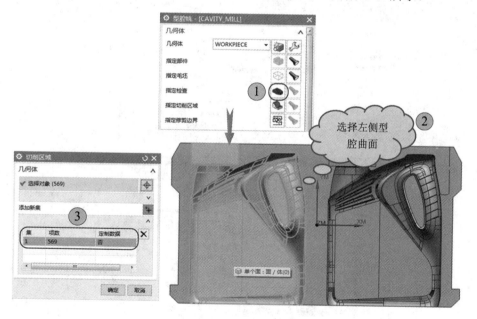

图 5-48 选择左侧型腔曲面为切削区域

（4）【刀轨设置】为【型腔铣】参数设置的主要内容。由于加工型腔曲面为开放轮廓，更适合使用【跟随部件】的切削方式加工，所以【切削模式】选项中设置为【跟随部件】

的方法。XY 方向步距设置【平面直径百分比】为 "75%"，Z 方向分层设置【最大距离】为每层 "0.5mm"。操作步骤如图 5-49 所示。

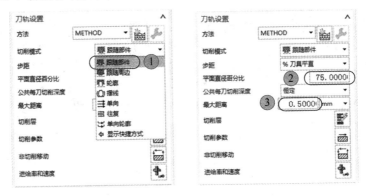

图 5-49　刀轨设置参数

(5) 单击 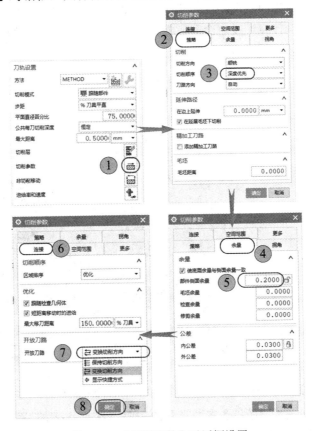【切削参数】图标，打开【切削参数】对话框。在【策略】标签中设置【切削顺序】为【深度优先】。选择【余量】标签，设置【部件侧面余量】为 "0.2mm"。选择【连接】标签，由于选择的是【跟随部件】的切削方式，在对话框中就会出现【开放刀路】的切削方法选择，在这里我们选择【变换切削方向】的方式。设置完成后单击【确定】按钮退出【切削参数】对话框，具体操作步骤如图 5-50 所示。

图 5-50　【切削参数】对话框设置

(6) 单击 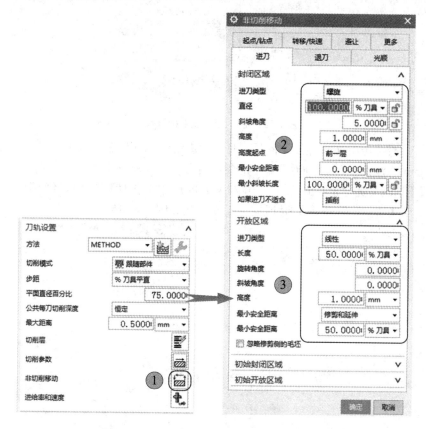 【非切削移动】图标，进入【非切削移动】对话框设置进刀参数。首先设置【封闭区域】下刀参数，型腔内【进刀类型】一般采用【螺旋下刀】的方式。由于面铣刀底面中心没有刀片，不存在切削能力，所以面铣刀是不能够直线下刀的，而且螺旋下刀的半径也要大才可以加工。因此我们在设置螺旋下刀时，设置下刀【直径】为刀具直径的"100%"，【斜坡角度】设置为"5 度"，【高度】设置为 1 mm，【最小斜坡长度】设置为100(注：面铣刀不能直线下刀，所以这个数值要填得较大，避免出现直线下刀路径，发生撞机事故)。【开放区域】设置进刀【长度】为刀具的"50%"，抬刀【高度】设置为"1 mm"，以减少抬刀距离。具体参数设置如图 5-51 所示。

图 5-51　选择【非切削移动】进刀参数设置

(7) 单击【非切削移动】对话框【转移/快速】标签设置快速抬刀高度。为提高加工速度减少抬刀高度，把【区域之间】和【区域内】的【转移类型】都改为【前一平面】，并且把抬刀【安全距离】都设置为"1mm"。这样做的好处是能减少不必要的抬刀，节约加工时间。但大部分快速移刀都是在工件零平面以下进行的，所以要求机床的 G00 运动方式必须是两点间的最短距离，不能是两点间的折线运动，否则会发生撞机事故。加工前一定要在MDI 下输入"G00 走斜线观察机床"的移动方式，如果错误，则需要修改机床参数，或者按照 NX 12.0【转移/快速】的初始设置，把【转移类型】全部设置为【安全距离-刀轴】。每次提刀都回到安全平面，避免发生碰撞。最后单击【确定】按钮退出【非切削移动】对话框，具体参数设置如图 5-52 所示。

图 5-52 【转移/快速】标签设置

(8) 单击 【进给率和速度】图标，打开【进给率和速度】对话框，设置【主轴速度】为"1000 rpm"(注：输入"1000"后一定要按后面的【计算器】图标，否则会报警)。输入【进给率】【切削】为"800 mmpm"，输入【更多】里【进刀】为"70%切削"，然后单击【确定】退出【进给率和速度】对话框，操作步骤如图 5-53 所示。

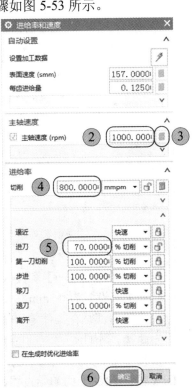

图 5-53 进给率和速度设置

(9) 单击 【生成】图标计算加工路径。单击【确定】退出型腔铣，如图 5-54 所示。

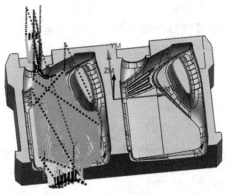

图 5-54　生成刀具路径图

5.5.2　二次开粗程序编制

因为第一次粗加工刀具直径较大，会有很多部位加工不到，所以需要选择直径小一点的刀进行二次开粗的加工。这里选用直径为 12 mm 的钻铣刀进行模具型腔的二次开粗工作。

(1) 点开【机床】视图，建立一把直径为 12 mm、下半径为 0.8 的钻铣刀，刀号为 4 号，刀具名称为 D12R0.8。

(2) 使用剩余铣的加工方法对模具型腔进行二次开粗编程，在【机床】视图中右键单击刚编好的【CAVITY_MILL】程序，选择【复制】选项，如图 5-55 所示。然后右键单击【D12R0.8】的钻铣刀，选择【内部粘贴】，如图 5-56 所示。形成一个新的提示错误的【型腔铣】程序，如图 5-57 所示。

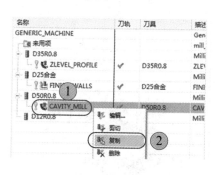
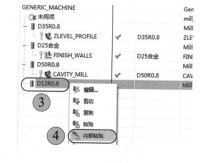

图 5-55　复制【型腔铣】程序　　　　图 5-56　内部粘贴【型腔铣】程序

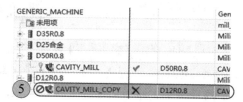

图 5-57　新的【型腔铣】程序

(3) 双击打开新复制过来的【型腔铣】程序，如图 5-57 圆圈 5 所示。设置 Z 向分层【最大距离】为 "0.2mm"，如图 5-58 所示。

(4) 单击【切削参数】图标，如图 5-59 所示，弹出【切削参数】对话框。

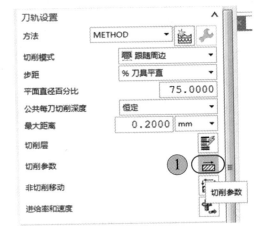

图 5-58 设置 Z 向分层为 "0.2 mm"　　　图 5-59 单击【切削参数】图标

(5) 在【切削参数】对话框中，选择【空间范围】标签，设置工件剩余毛坯。在【过程工件】下拉菜单中选择【使用基于层的】选项，设置本次加工毛坯为上次加工剩余的部分。在【剩余铣】中此选项默认为【使用基于层的】选项，【型腔铣】和【剩余铣】的不同只有这一个选项。然后单击【确定】按钮，如图 5-60 所示。

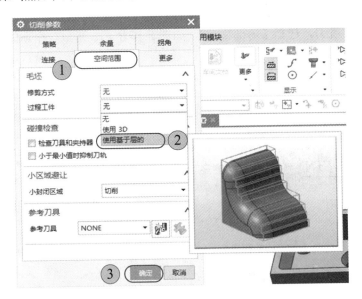

图 5-60 设置【空间范围】标签

(6) 单击 ⬆️【进给率和速度】图标，设置【主轴速度】为 "4000 rpm"，单击转速右侧 🔲【计算器】图标，最后单击【确定】按钮退出【进给率和速度】对话框，如图 5-61 所示。

(7) 单击 ▶️【生成】图标计算加工程序，生成【型腔铣】加工路径，单击【确定】按钮退出【型腔铣】设置，如图 5-62 所示。

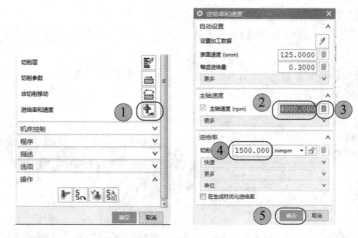

图 5-61　设置【进给率和速度】

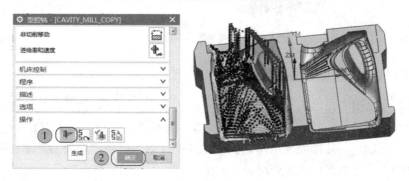

图 5-62　生成二次开粗刀具路径

（8）使用 R5 球刀粗加工 D12R0.8 钻铣刀剩余部分。点开【机床】视图，建立一把直径为 10mm 的球刀，刀号为 5 号，刀具名称为 R5。

（9）在【机床】视图中右键单击上一步【剩余铣】程序，选择【复制】选项，然后右键单击【R5】刀具，选择【内部粘贴】。形成一个新的提示错误的【型腔铣】程序，具体操作方法参考如图 5-57 所示。

（10）双击打开新复制的【型腔铣】程序。【步距】选择为【恒定】，XY 步距【最大距离】数值设置为"0.3mm"，设置 Z 向分层【最大距离】为"0.2mm"，如图 5-63 所示。

图 5-63　设置 XY 步距和 Z 向分层

(11) 单击 💺【进给率和速度】图标，设置【主轴速度】为"8000 rpm"，单击转速右侧 ▣【计算器】图标，最后单击【确定】按钮退出【进给率和速度】对话框，具体操作方法参考图 5-61。

(12) 单击 ⬅【生成】图标计算加工程序，生成【型腔铣】加工路径，单击【确定】按钮退出【型腔铣】设置，如图 5-64 所示。

图 5-64　生成二次开粗刀具路径

5.5.3　曲面精加工程序编制

曲面精加工常用的程序为【固定轴轮廓铣】，在新版本的 NX 软件中能够实现按照曲面倾斜的不同角度设置出不同的加工方法，也就是 NX 里的【陡峭和非陡峭】。一般编辑精加工曲面程序的原则是，陡峭面使用等高的形式从上往下一层一层地加工，非陡峭面使用往复或者环绕的方式精加工，下面介绍一下曲面精加工程序的编制。

(1) 单击【创建工序】图标，弹出【创建工序】对话框，在【类型】下拉菜单中选择【mill_contour】曲面铣选项。选择 ◆【区域轮廓铣】图标，在【程序】下拉菜单中选择【2】程序组，【刀具】下拉菜单中选择【R5】球刀，【几何体】下拉菜单中选择【WORKPIECE】，然后单击【确定】按钮，操作步骤如图 5-65 所示。

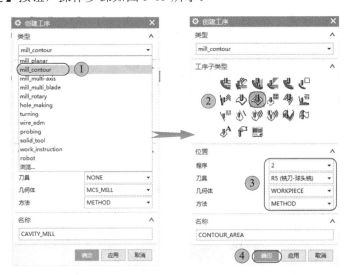

图 5-65　创建【区域轮廓铣】工序

(2) 在弹出的【区域轮廓铣】对话框中左键选择 📦【指定切削区域】图标，选择模具左侧单一型腔曲面，单击【确定】退出【切削区域】对话框，如图 5-66 所示。

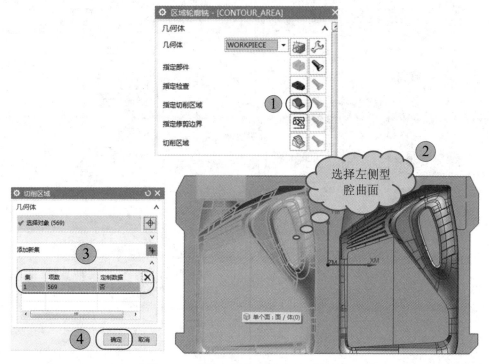

图 5-66　选择左侧型腔曲面为切削区域

(3) 在【驱动方法】选项中单击右侧 🔧【编辑】图标，打开【区域铣削驱动方法】对话框。在【陡峭空间范围】中，【方法】选项中选择【陡峭和非陡峭】。设置刀具路径的走刀方式为"65～90"度的曲面加工范围，使用等高的形式加工。0～65 度的曲面加工范围，使用往复的形式加工。然后设置【非陡峭切削模式】为【往复】，【步距】改为【恒定】，每刀间的【最大距离】为"0.15mm"。(注意：精加工曲面时步距过大会使加工出的曲面光洁度差，步距过小会使曲面光洁度好但加工时间长，经过多年的加工经验得出 0.15mm 的步距加工出的曲面效果最好)。【步距已应用】选择【在部件上】，这样加工出来的曲面效果好。【陡峭切削模式】选择【往复深度加工】，【深度切削层】选择为【恒定】，【深度加工每刀切削深度】【合并距离】【最小切削长度】三个选项都设置成为"0.15mm"。然后单击【确定】按钮退出设置。具体设置方法如图 5-67 所示。

(4) 单击 🎛【切削参数】图标，打开【切削参数】对话框。单击【余量】标签，设置【内公差】【外公差】都为"0.01mm"，然后单击【确定】退出【切削参数】对话框，如图 5-68 所示。

(5) 单击 🎛【进给率和速度】图标，打开【进给率和速度】对话框，设置【主轴转速】为"10000rpm"。输入【进给率】【切削】为"1500mmpm"，设置【更多】里【进刀】为"70% 切削"，然后单击【确定】退出【进给率和速度】对话框，设置方法参考图 5-61。

(6) 单击 🏳【生成】图标，计算加工路径。单击【确定】退出区域轮廓铣，如图 5-69 所示。

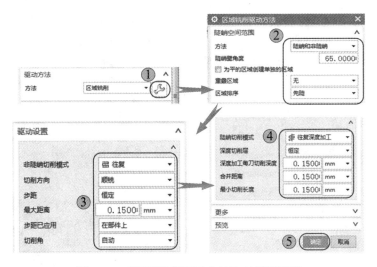

图 5-67 【区域铣削驱动方法】对话框设置

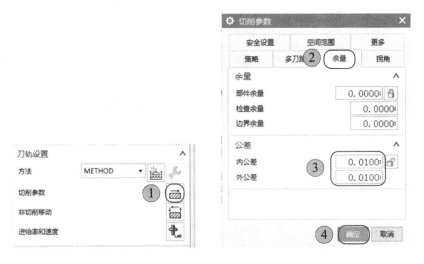

图 5-68 设置余量【内公差】【外公差】均为"0.01 mm"

 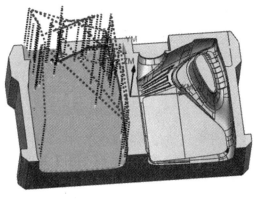

图 5-69 生成刀具路径图

5.5.4　精加工曲面清角程序编制

使用 R3 球刀编辑曲面清角程序，曲面清角程序的走刀方法和精加工的方法类似，也是按角度的方式分为陡峭和非陡峭加工，下面是曲面清角程序的设置方法。

(1) 点开【机床】视图，建立一把直径6 mm、长100 mm的球刀，刀号为6号，刀具名称为R3。

(2) 使用清根铣完成曲面小圆角处的精加工，在【机床】视图中右键单击新编的【CONTOUR_AREA】程序，选择【复制】选项。然后右键单击【R3】刀具，选择【内部粘贴】。形成一个新的提示错误的【区域轮廓铣】程序，如图 5-70 所示。

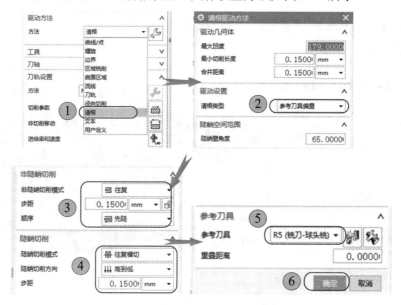

图 5-70　复制出新的【区域轮廓铣】程序

(3) 单击【驱动方法】【方法】下拉菜单，选择【清根】的方法，打开【清根驱动方法】对话框。在【驱动设置】【清根类型】下拉菜单中选择【参考刀具偏置】，参考刀具偏置是指按照程序里选定的参考刀具的直径，自动计算出剩余圆角部分进行清角加工。在【非陡峭切削模式】下拉菜单中选择【往复】的切削模式，【步距】设置为"0.15mm"，加工【顺序】为【先陡】，就是先加工陡峭面然后再加工非陡峭面，这种方法有助于延长刀具寿命，以避免发生碰撞折刀的问题。【陡峭切削模式】选择【往复横切】，往复横切的路径和深度轮廓铣的加工原理类似，都是从上往下一层一层地加工。【陡峭切削方向】选择【高到低】，【步距】设置为"0.15mm"。最后打开【参考刀具】，设置【参考刀具】为上一把铣刀【R5(铣刀-球头铣)】。单击【确定】退出清根设置，具体设置方法如图 5-71 所示。

图 5-71　【清根驱动方法】对话框设置

(4) 单击 【生成】图标计算加工路径。单击【确定】退出【区域轮廓铣】，如图 5-72 所示。

图 5-72　生成刀具路径图

5.6　变换刀具路径制作另一型腔程序

使用平移刀具路径的方法制作出右侧型腔的加工程序，具体操作方法如下所示。

(1) 单击 【程序顺序】视图图标，使【工序导航器】切换到【程序顺序】视图。单击屏幕左上角 【创建程序】图标，弹出【创建程序】对话框，输入名称"3"，单击【确定】弹出【程序】对话框，再单击【确定】按键退出【程序】对话框。在【程序顺序】视图导航器下增加一个【3】的程序组，如图 5-73 所示。

图 5-73　建立新的【3】程序组

(2) 单击【测量】图标，【投影距离】选项选择坐标系 X 方向，在两个模型型腔上分别选择两个相同位置的端点，测得两型腔中心距为 250mm，如图 5-74 圆圈 2 所示。

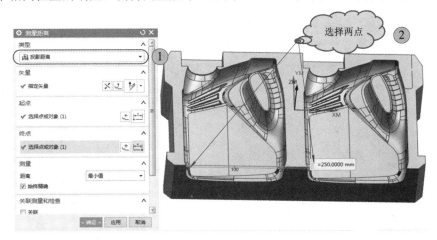

图 5-74　测量两型腔中心距为 250mm

（3）全选【2】程序组下所有的加工程序，如图 5-75 所示。

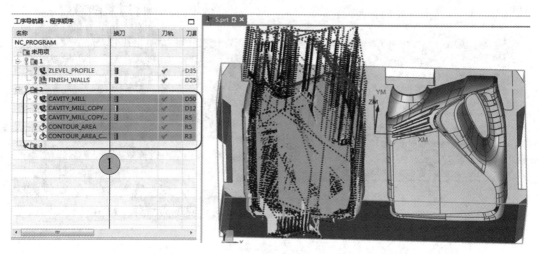

图 5-75　全选【2】程序组下所有程序

（4）右键选择【对象】，然后单击【变换】选项，如图 5-76 所示，弹出【变换】对话框。

（5）在【类型】下拉菜单中选择【平移】指令，在【变换参数】【XC 增量】选项中输入中心距"250mm"，【结果】选择【复制】。单击【确定】退出，如图 5-77 所示。

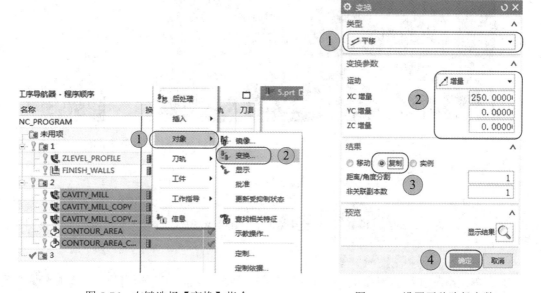

图 5-76　右键选择【变换】指令　　　　图 5-77　设置平移路径参数

（6）左键拖动新生成的右侧型腔程序到【3】程序组下，如图 5-78 所示。

（7）仿真编辑完所有加工程序，单击左上方 █🗗【程序顺序】视图图标，在【程序顺序】视图界面全选所有编完的加工程序。单击【主页】下【确认刀轨】图标，打开【刀轨可视化】对话框。单击【3D 动态】，切换模拟动画为三维立体模型，如图 5-79 所示，然后单击下方 ▶【播放】图标，完成工件底面路径的加工仿真。

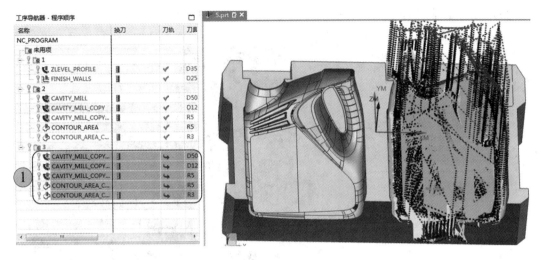

图 5-78　生成右侧型腔加工程序

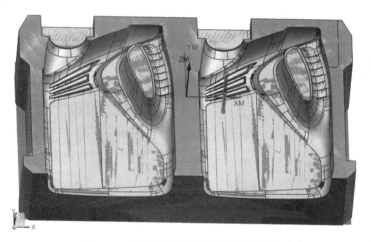

图 5-79　第二序编完程序的加工仿真　　　　　　　　项目五第二序机床仿真

5.7　生成 G 代码文件

(1) 按住电脑键盘【Ctrl】键，鼠标左键单击选择【1】程序组下所有程序，单击![后处理]【后处理】图标，弹出【后处理】对话框，如图 5-80 所示。

(2) 在【后处理器】选项中选择【FANUC0i】后处理文件，单击【输出文件】选项下的![浏览]【浏览以查找输出文件】图标，弹出【指定 NC 输出】对话框，(首先在 D 盘创建【nc】文件夹)选择 D:\nc 目录，输入文件名为"1"，单击【OK】返回【后处理】对话框。确定文件名位置为 D:\nc\1，文件扩展名为.nc。单击【确定】退出设置，如图 5-81所示。

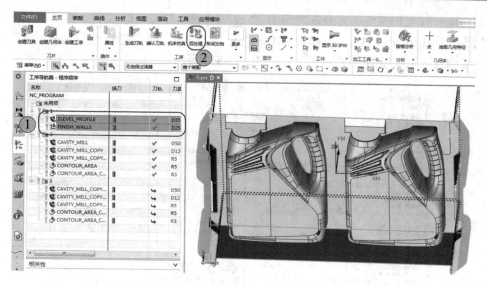

图 5-80　选择【1】程序组下所有程序，单击【后处理】图标

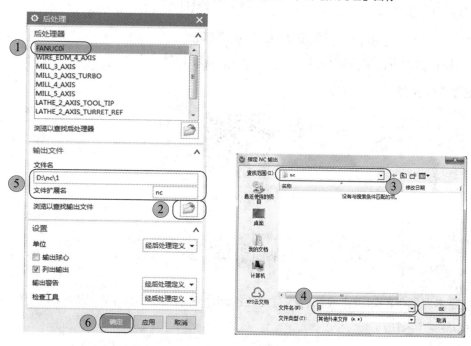

图 5-81　设置后处理文件位置及名称

(3) 单击【确定】后弹出【多重选择警告】对话框，单击【确定】按键将所有程序输出在一个程序组下显示，如图 5-82 所示。

图 5-82　弹出【多重选择警告】对话框

(4) 弹出 G 代码文件，并且在电脑 D 盘 nc 文件夹下生成 1.nc 文件，生成 G 代码文件后，【1】程序组下所有程序前面都显示出绿色对勾，表示已经出完 G 代码文件，未出 G 代码文件的加工程序前显示为黄色感叹号，如图 5-83 所示。

图 5-83　生成 G 代码文件

(5) 以相同的方法生成其他两个程序组的加工程序。

编程操作视频

项目五编程操作视频

课后练习

按照本课所学的知识完成课后练习文件的程序编制，课后练习见图 5-84 和光盘"练习"文件夹中 5.prt 文件。

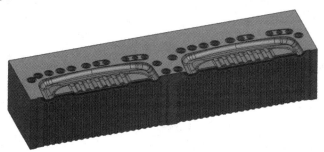

图 5-84　项目五课后练习图

附录 重点项目英文对照

Project I Programming for the Machining of Switch Box Frame

➤ Case Description ✍

This project takes the machining programming of the switch box frame as an example, which explains the processing technology of switch box frame, the selection of processing method, the comparison of different machining methods used in the same machining operation, the selection of cutting tools and the cautions of planar machining programming.

➤ Learning Objectives ✍

Through learning the processing programming of the parts of switch box frame, readers can understand and master the processing programming methods of three-axis planar parts using NX software, and draw inferences by analogy.

➤ Learning Tasks ✍

Planar contour parts are commonly used parts in machining. There are many types of machining, and most of them are carried out in batch form, which are often used in electronic, medical, mechanical, aviation and other industrial fields.

This case takes the box processing programming in the switch box as an example to explain the three-axis plane programming of NX software. In this process, the emphasis is laid on learning and understanding the machining methods and applications of planar contour programs of NX software, and machining programs can be generated depending on different methods and processes.

Figure 1-1 is the flowchart of NX software machining process, which illustrates the process and steps of creating and processing programs. This flowchart will be used as a guide throughout this textbook.

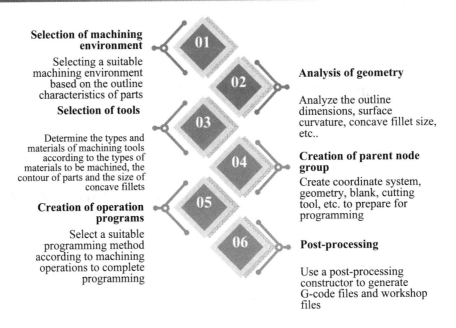

Selection of machining environment

Selecting a suitable machining environment based on the outline characteristics of parts

Selection of tools

Determine the types and materials of machining tools according to the types of materials to be machined, the contour of parts and the size of concave fillets

Creation of operation programs

Select a suitable programming method according to machining operations to complete programming

Analysis of geometry

Analyze the outline dimensions, surface curvature, concave fillet size, etc..

Creation of parent node group

Create coordinate system, geometry, blank, cutting tool, etc. to prepare for programming

Post-processing

Use a post-processing constructor to generate G-code files and workshop files

Figure 1-1   Flowchart of UG NX 12.0 Software Machining Process

1.1   Machining process specification of switch box frame

The machining process specification is to describe each step of machining process, generally including the area to be machined, machining type (planar milling, surface milling, hole processing, etc.), process content description, parts clamping, required tools and other information necessary to complete the machining.

The outline dimension of the switch box frame is 80 mm × 40 mm × 40 mm rectangle, and the material is aluminum alloy 6061. The product is machined in batches. In order to facilitate batch machining and reduce machining costs, blanks with slightly larger outline dimensions are usually selected to machine the required outline dimensions when purchasing the raw materials of aluminum products. It is unnecessary to customize the size of the whole six surface finishing rough material. Therefore, blanks with 5mm margins on both sides and rectangular aluminum alloy profiles with dimensions of 85 mm × 45 mm × 45 mm are selected in this project. The following process briefly describes the machining process of switch box frame, as shown in Figure 1-1. It includes process number, sequence number, machining tool, process content, process name and tool name. Their definitions are as follows:

Process number—Sequence number of process distribution.

Sequence number—Sequence number of machining under one of the processes.

Machining tool—Selection of machining tool for an operation.

Process content—Description of detailed operation content.

Process name—Actual name of a particular feature or task of a machining operation.

Tool name—Type and material of the tool used for machining.

1.1.1 Case on process analysis

Figure 1-2 shows the model of switch box frame. This workpiece needs to be machined in both up and down directions and the parts need to be clamped twice to complete the manufacturing of the workpiece. In order to ensure the dimensional accuracy and form and position tolerance of machined parts, it is necessary to determine the machining operation scheme of the workpiece before programming.

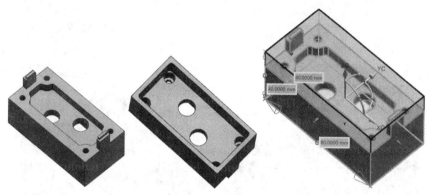

Figure 1-2   Model of Switch Box Frame

Scheme I:

Clamp the front surface of the part to be machined first in a depth range of 0—14.5 mm using a vise, and then clamp the two raised portions milled and the lower half, as shown in Figure 1-3. When machining the floor, because the clamping surface is small and the material is soft aluminum alloy, the workpiece is likely to fly out when the clamping force of the vice is too small, but if the force is too intensive, the workpiece will deform easily and vibrate violently during machining, resulting in oversize of workpiece, a great difficulty in machining and high rejection rate, which are unusable.

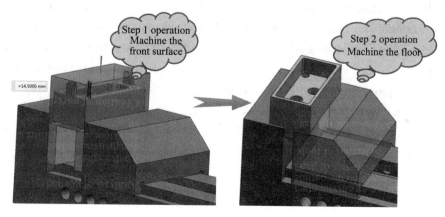

Figure 1-3   Diagram of Machining Scheme I

Scheme II:

Clamp the floor of the part to be machined first with a depth of 10 mm using a vice, machine all parts including floor outline, cavity, hole, etc., which can be machined except the two bosses on the front surface, and then clamp the milled rectangular part and clamp the two bosses and cavity on the front surface of the part, as shown in Figure 1-4. If this scheme is used, the clamping area of the workpiece will be large and the stock of blanks to be machined will be small in Step 2 operation of machining. In addition, there will not have too many requirements for the clamping force of parts, so the clamping speed of parts and the qualification rate of machined parts can be improved. Therefore, scheme II is selected.

Figure 1-4   Diagram of Machining Scheme II

1.1.2   Case on selection of machining tools

In this example, the machining material of the parts of switch box frame is aluminum alloy 6061. The outline dimensions are 80mm × 40 mm × 40 mm, and the minimum radius of concave fillet is 2 mm, as shown in Figure 1-5. The inner and outer contours of the cavity are composed of planes. Roughing is conducted for the cavity at first. A large amount of material needs to be removed during roughing, and the cutting tool will bear a great force. Therefore, D16R0.4 inserted drilling and milling tools with large diameter shall be selected. In this way, the tools are not easy to vibrate, break and snap. A $\phi 4$ cemented carbide cutting tool for aluminum is used to perform reroughing, and the residual fillet stock after roughing is removed. Then a $\phi 10$ cemented carbide cutting tool for aluminum is used for finish milling of the contour and floor of the part. Finally, proper center drill and drill tool are selected according to the size of the hole diameter of the part to machine all the holes on the part.

The machining process of parts will be arranged rationally according to the machining process scheme of parts and the selection of cutting tools as analyzed above. The processing technology for this case will be determined by the following principle of rough processing first and fine processing later, surface processing first and hole processing later with benchmark unification, as shown in Table 1-1.

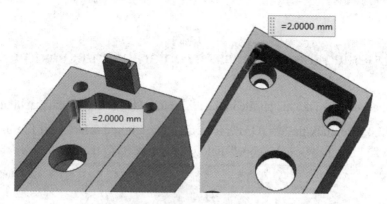

Figure 1-5　Diagram of Minimum Concave Fillet

Table 1-1　Process Sheet of Switch Box Frame

Process number	Sequence number	Machining tool	Process content	Process name	Tool name
Floor	1	Machining center	Roughing	Cavity drill	D16R0.4
Floor	2	Machining center	Reroughing	Rest milling	D6
Floor	3	Machining center	Floor finishing	Floor wall milling	D10
Floor	4	Machining center	Side wall finishing	Floor wall milling	D6
Floor	5	Machining center	Side wall finishing	Finish milling of wall	D10
Floor	6	Machining center	Center drilling	Spot drill	Center drill D6
Floor	7	Machining center	Machining of $\phi 4.5$ hole	Drilling	Drill D4.5
Floor	8	Machining center	Machining of $\phi 8$ counterbore	Drilling	D8
Floor	9	Machining center	Machining of $\phi 13$ hole	Drilling	Drill D13
Upper surface	10	Machining center	Roughing	Cavity drill	D16R0.4
Upper surface	11	Machining center	Roughing	Cavity drill	D10
Upper surface	12	Machining center	Reroughing	Rest milling	D4
Upper surface	13	Machining center	Floor finishing	Floor wall milling	D10
Upper surface	14	Machining center	Side wall finishing	Floor wall milling	D6
Upper surface	15	Machining center	Angle-clearing	Floor wall milling with IPW	D4

1.2 Opening the model file to enter the machining module

Open the CD attached, open the model file in the example folder, and enter the machining module.

(1) Start NX 12.0, and click the ✎ [Open] icon at upper left and select 1.prt file in "the sample folder" in the CD from the "Search Scope" dialog box, Click [OK] button as shown in Figure 1-6.

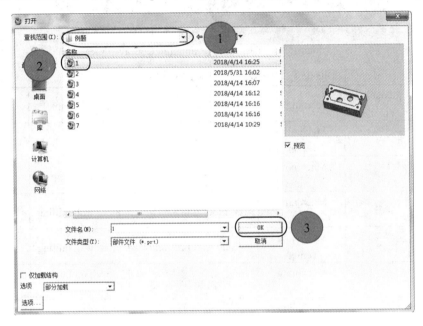

Figure 1-6　Open 1.prt File in the "Sample" Folder on CD

(2) Click [Application] button, then click [Machining] (or press the keyboard shortcut Ctrl + Alt + M directly) to enter the NX12.0 "machining" module, as shown in Figure 1-7.

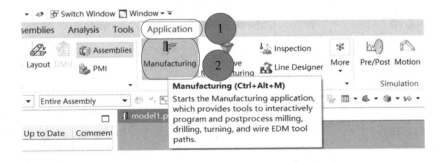

Figure 1-7　Enter the UG NX12.0 "Machining" Module

(3) Select the [CAM session configuration] [cam_general] by default. In the [CAM assembly] dialog box created, select [mill_planar], and click [OK] to enter the planar milling interface, as shown in Figure 1-8.

Figure 1-8　Setting of [Machining Environment] Dialog Box

1.3　Creation of parent node group

The parent node group includes geometry view, machine tool view, program order view, and machining method view.

(1) Geometry view: direction and clearance planes of the "machining coordinate system" can be defined, and parameters such as "parts", "blank" and "check" geometry can be set.

(2) Machine tool view: the cutting tools can be defined, you can specify the milling cutter, drill bit, turning tool etc., and save the tool-related data as the default value of the corresponding post-processing command.

(3) Program order view: it can arrange completed programs in the folder by groups, and arrange machining programs in order from top to bottom.

(4) Machining method view: it is used to define the type of cutting methods (roughing, finishing, semi-finishing). For example, parameters such as "internal tolerance", "external tolerance" and "part stock" are set here.

1.3.1　Creation of machining coordinate system

The proedures for creating the machining coordinate system in the [geometry view] menu are as follows:

(1) Switch the [Operation Navigator] to the [Geometry], as shown in Figure 1-9.

(2) Double-click [MCS_Mill] icon to enter [MCS Mill] dialog box, as shown in Figure 1-10.

Figure 1-9　Switching of [Geometry View]　　　Figure 1-10　[MCS Mill] Dialog Box

(3) Since the blank is not machined, the surface shape and dimensions may be irregular. To ensure a uniform machining stock on each face, it is better to place the coordinate system in the middle of the blank. Click the ⬚ [Coordinate System Dialog Box], as shown in Figure 1-11, circle 1. In the [Coordinate System] dialog box that appears, select [CSYS of Object], as shown in Figure 1-12, circle 2. It automatically sets the coordinates to the center point of the selected plane. Click the box on the upper surface of the workpiece to automatically capture the coordinate position at the center point on the upper surface of the workpiece, as shown in Figure 1-13, circle 3. If there is no interference in the clearance plane, the default state [Automatic Plane] can be selected, and the safe clearance Distance is "10 mm". If there is interference, the safe clearance distance can be set to"50 mm~100 mm". At last, click [OK] to exit the [MCS Mill] dialog box, as shown in Figure 1-13, circles 4 and 5.

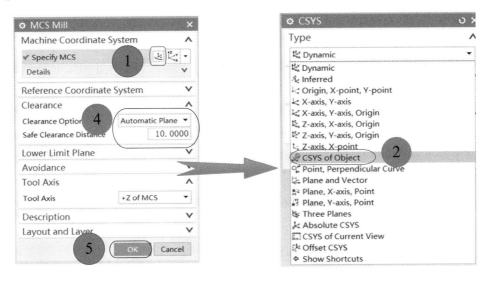

Figure 1-11　Select [Coordinate System Dialog Box]　　　Figure 1-12　Select [Coordination System of Object]

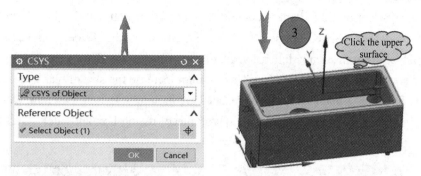

Figure 1-13　Click the Box on the Upper Surface of the Workpiece and Set the Coordinate System

1.3.2　Creation of the Geometry of Machining Part

The operation steps for creating the geometry of machining part, part blank, and checking geometry in the [geometry] view menu are as follows:

(1) Double-click WORKPIECE to open the [Workpiece] dialog box, as shown in Figure 1-14.

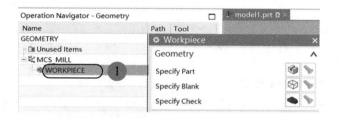

Figure 1-14　Open the [Workpiece] dialog box

(2) Click [Select or Edit Part Geometry] as shown in Figure 1-15 circle 1, and the [Part Geometry] dialog box pops up. Click the processed part to make it orange. Next, click [OK] to exit the [Part Geometry] dialog box, as shown in Figure 1-15 circle 2.

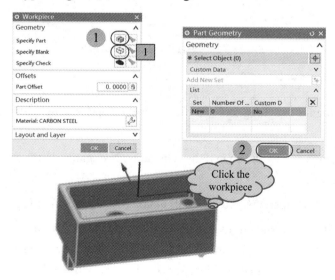

Figure 1-15　Establish Part Geometry

(3) Click ⊕ [Select or Edit Blank Geometry], as shown in Figure 1-15, box 1. In the [Blank Geometry] dialog box popped up, select [Bounding Block] in the [Type] options, as shown in Figure 1-16, box 2. Set the blank dimension as follows: increase the size on one side by "2.5 mm", increase the size of blank floor by "5 mm", and do not increase the size of blank top, as shown in Figure 1-16, box 3. The blank dimensions reached are 85 mm × 45 mm × 45 mm. Click [OK], as shown in Figure 1-16, box 4. Return to the [Workpiece] dialog box and click [OK] to exit and complete setting, as shown in Figure 1-16, box 5.

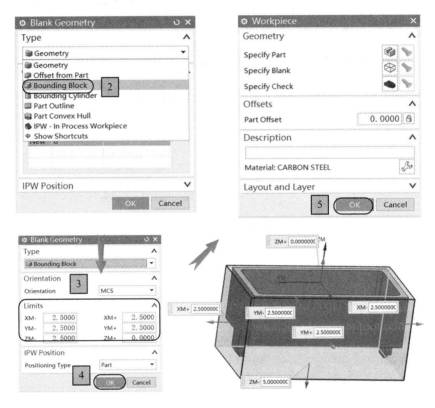

Figure 1-16   Establish Blank Geometry

1.3.3   Creation of Tool

The steps for creating the roughing tool—D16R0.4 drilling and milling tool in the [machine tool] view are as follows:

(1) Click ⚙ [Machine View] and switch the [Operation Navigator] to the [Machine Tool] View, as shown in Figure 1-17.

(2) Click [Create Tool] to open the [Create Tool] dialog box, as shown in Figure 1-18.

(3) Select 𝕴 [MILL] as shown in Figure 1-19 circle 1 to create a flat-end tool. Enter "D16R0.4" (representing the inserted drilling and milling cutter with a diameter of 16 mm and a fillet radius of 0.4 mm) in the name position, as shown in Figure 1-19.

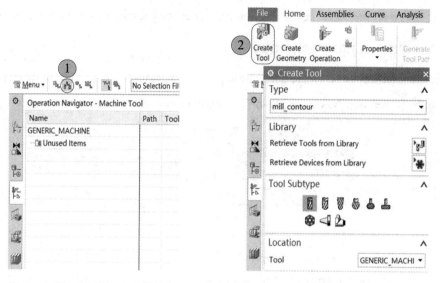

Figure 1-17　[Machine Tool] View　　　　Figure 1-18　Dialog Box for [Creating Tool]

(4) Set the [Diameter] to "16", the [Lower Radius] set to "0.4", and the [Tool Number], [Adjust Register], [Cutcom Register] to "1" (this value represents the number of tool, tool radius compensation, and tool length compensation. In order to avoid collision, it is preferable to use the same number). Click [OK] to complete tool setup, as shown in Figure 1-20.

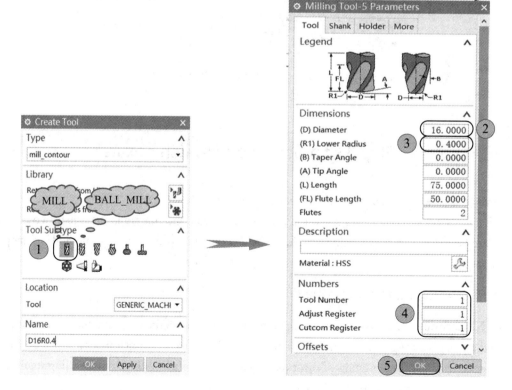

Figure 1-19　Create a Flat-end Tool　　　　　　　Figure 1-20　Set the Flat-end Tool

Other tools can be set separately based on given parameters before editing the machining program.

1.3.4 Creation of program group

In the [Program Order] view, create a machining program group folder as the follow steps:

(1) Switch the [Operation Navigator] to the [Program Order] view, as shown in Figure 1-21.

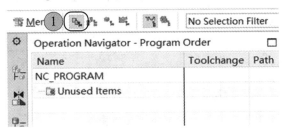

Figure 1-21　Switch the [Operation Navigator] to the [Program Order] view

(2) Double-click the text of [PROGRAM] program group to change the file name to [Bottom] (or right-click [PROGRAM] program group and select [Rename] to change the name), as shown in Figure 1-22, circle 2.

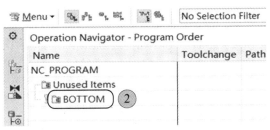

Figure 1-22　Double-click to Modify the Program Group Name

(3) Save the files.

1.4　Step 1 of machining of switch box frame

According to the previous machining analysis, the step 1 machining of the workpiece is to machine the inner and outer cavities of the part upwards from the floor, and the upper two convex parts are left for step 2 machining. The following is the programming method of step 1 machining.

1.4.1 Programming for roughing

When NX software conducts programming on the basis of model diagram and blank, the simplest and most effective roughing program is the cavity mill in curved surface machining. The cavity mill can be used for the roughing of most parts. The steps for using the cavity milling method to complete the programming for roughing of parts are as follows:

In order to better reflect the order of machining programs, we will use the [Program Order] viewto complete all programming.

(1) Click [Create Operation] to open the [Create Operation] dialog box, as shown in Figure 1-23, circle 1.

(2) In the [Type] pull-down menu, select [mill _ contour], as shown in Figure 1-24, circle 2. Click ⬚ [Cavity Mill] icon, select the created [Bottom] program group from the [Program] pull-down menu, select the inserted drilling and milling tool [D16R0.4] from the tool pull-down menu, and select the created [WORKPIECE] geometry from the geometry pull-down menu. The user can enter a program name as per machining requirements in the name column. The name will not be specially modified here, and will be filled in according to the default name. After that, click [OK], as shown in Figure 1-25, circles 4 and 5.

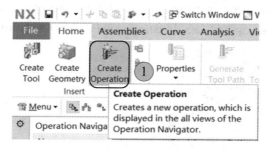

Figure 1-23　Click [Create Operation]

Figure 1-24　Select [mill-contour] Option　　Figure 1-25　Create a [Cavity Mill] Operation

(3) In the [Cavity Mill] dialog box that appears, if [WORKPIECE] is correctly set in the [Geometry] view, the [Specified Part] and [Specified Blank] options should appear gray after

entering the [Cavity Mill]. The  [Display] icon on the right is colored, and click it to display the selected geometry. If parts and blanks can be selected after entering [Cavity Mill], it means [WORKPIECE] in the [Geometry] view is not set, or the geometry is not selected as [WORKPIECE] before entering the program. The options [Specify Cut Area], [Specify Check] and [Specify Trim Boundaries] generally do not requiring setting in cavity mill programming as in this case, and will be set according to part machining requirements, as shown in Figure 1-26.

(4) Click the drop-down arrow in the [Tool] menu to show that the selected machining [Tool] is D16R0.4 inserted drilling and milling tool (note: This work can be ignored in case of correct selection in previous steps), as shown in Figure 1-27.

(5) Click the drop-down arrow in the [Tool Axis] menu to show the default tool axis [+ZM Axis]. The tool axis of 3-axis machining center generally uses +ZM axis. This item will be modified only when a multi-axis machine tool is used. (Note: This option does not have to be selected during 3-axis machining programming, and default settings will be used), as shown in Figure 1-27.

Figure 1-26　Dialog Box for Geometry Setting

Figure 1-27　Options of Tool and Tool Axis

(6) [Tool Axis Settings] is the main content of the parameter settings of cavity mill. Two patterns are generally used in cutting pattern options: Follow Part and Follow Periphery. The [Follow Part] pattern is suitable for machining workpieces with open contour, allowing the tool to machine from outside to inside and engage from the outside of the workpiece. The [Follow Periphery] pattern is more suitable for machining workpieces with closed contour, allowing the tool to machine from inside to outside and reducing the change of engage position during cavity machining. In this case, there are many machining faces with closed contour at the floor, so the machining pattern [Follow Periphery] is selected, as shown in Figure 1-28, circle 1.

(7) In general, 70% to 80% of the tool diameter is used for the tool stepover in the XY direction of roughing, and 50% or less of the tool diameter is used for finishing. 80% to 100% stepover in the XY direction is not recommended generally because a too large tool stepover is equivalent to full tool cutting in each cut and will affect the tool life and machine tool precision due to an excessive tool stressing. Moreover, if the stepover is too large, the floor machined will have a poor finish and may have tool marks, so the value is set to "75" in the [Percent of Flat Diameter]

option, as shown in Figure 1-29, circle 2.

(8) For roughing using a D16R0.4 drilling and milling tool, the stepover in Z-direction is generally set to "0.3 mm~0.7 mm" per level. The [Maximum Distance] option is to set the stepover of each level in Z-direction of the tool, so it is set to an intermediate value of "0.5 mm" per level, as shown in Figure 1-29, circle 3.

(9) Click ⚒ [Cutting Parameters] to open the [Cutting Parameters] dialog box, as shown in Figure 1-29, circle 4.

Figure 1-28　Select [Follow Periphery]

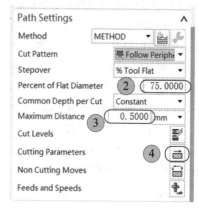

Figure 1-29　Stepover Settings in XYZ Directions

(10) Set the [Cut Order] as [Depth First] in the [Strategy] label. In this mode, the tool will machine from top to bottom depending on different areas, therefore reducing the tool lifting and passing path and reducing the machining time. If [Level First] is selected, the tool will move across different areas for machining at the same depth, increasing a lot of tool lifting paths and the machining time. In general, preference is given to [Depth First], as shown in Figure 1-30.

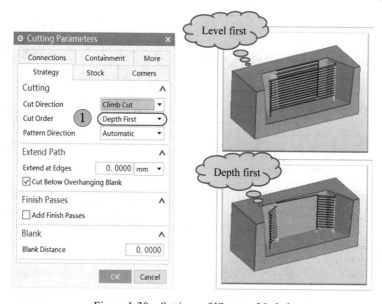

Figure 1-30　Settings of [Strategy] Label

(11) Select the [Stock] Label to set [Part Side Stock] to "0.2 mm" (note: The stock of roughing tool is typically set to "0.2 mm"), and click [OK] to exit the [Cutting Parameters] dialog box, as shown in Figure 1-31.

(12) Click the 📇 [Non Cutting Moves] icon as shown in Figure 1-32, circle 1, to enter the [Non Cutting Moves] dialog box, which is shown in Figure 1-33.

(13) In order to set engage parameters, first set the engage parameters in the [Closed Area]. The tool in the cavity is generally engaged in a [Helical] form. Principles of helical engagement: When the cavity size can meet the requirement of helical engagement, the method of helical engagement can be selected. When the cavity size fails to meet the requirement of helical

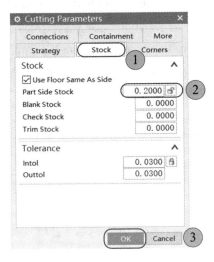

Figure 1-31　Settings of [Stock] Label

engagement, the method of inclined or ramping engagement is selected. The [Diameter] of helical engagement is set to 50% of the tool diameter, the [Ramp Angle] to "5 degrees", the [Height] to "1 mm", and the [Minimum Ramp Length] to "0" (Note: If the machining tool can be engaged linearly, the number can be input as 0. If the tool cannot be engaged linearly, the number shall not be less than 50; otherwise collision will occur), as shown in Figure 1-33, circle 2. In the [Open Area], the engagement [Length] is set to "50%" of the tool length, and the lifting [Height] is set to "1 mm" to reduce the lifting distance. Specific parameter settings are shown in Figure 1-33, circle 3.

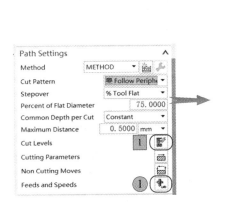

Figure 1-32　Select Non Cutting Moves

Figure 1-33　Engage Parameter Settings

(14) Click the [Transfer/Rapid] label to set the quick tool lift height. To increase the machining speed and reduce the lifting height, the lifting height both [Between Regions] and [Within Regions] are changed to [Previous Plane], and the [Safe Clearance Distance] are set to "1 mm" as shown in Figure 1-34, circle 1. In this way, unnecessary tool lifting can be avoided and the machining time is reduced. However, most of the quick tool moves occur below the zero plane of the workpiece, so the G00 motion of the machine tool must be a linear motion between two points rather than a broken line motion between two points; otherwise a collision accident may occur. Before machining, be sure to input the ramping mode for "G00 under MDI" to observe the movement mode of the machine tool. If it is not correct, it is necessary to modify the machine tool parameters. Or it is acceptable to set [Transfer Type] to [Safe Clearance Distance - Tool Axis] according to the initial settings of NX 12.0 [Transfer/Rapid]. At last, click [OK] to exit the [Non Cutting Moves] dialog box. Specific parameter settings are shown in Figure 1-34.

Figure 1-34　Settings of Transfer/Rapid Label

(15) Click 📝 [Cutting Levels] icon to enter the [Cutting Levels] dialog box, as shown in Figure 1-32, box 1.

(16) In the [Cutting Levels] dialog box, click ✖ several times to delete all the values from the list, as shown in Figure 1-35.

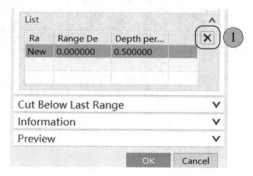

Figure 1-35　Delete List Values

(17) Click [Select Object] in [Range Definition] to select the part, as shown in Figure 1-36, circle1. Set the [Machining Depth] to "28 mm". Click [OK] to exit [Cutting Levels] dialog box.

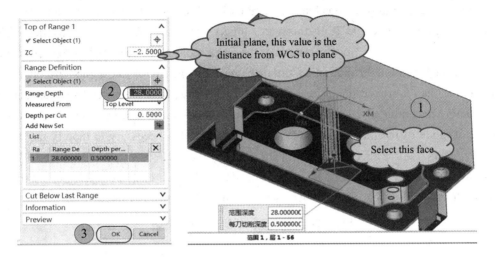

Figure 1-36   Select Cutting Depth

(18) Click [Feeds and Speeds] icon to open the [Feeds and Speeds] dialog box and set the [Spindle Speed] to "2500 rpm" (note: be sure to press the following [Calculator] icon after value "2500 rpm" is input; otherwise the system will alarm). The [Cut] option in in [Feed Rates] is set as "1500 mmpm", and the [Engage] option in [More] is set to "70% cut" rate. Next, click [OK] to exit the [Feeds and Speeds] dialog box, as shown in Figure 1-37.

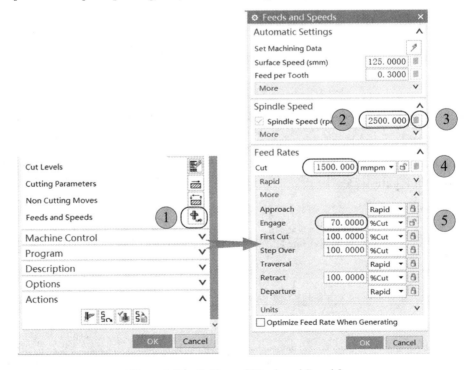

Figure 1-37   Settings of [Feeds and Speeds]

(19) Click  [Generate] icon to calculate the machining path, as shown in Figure 1-38, circle 1. Click [OK] to exit [Cavity Mill], as shown in Figure 1-38, circle 2.

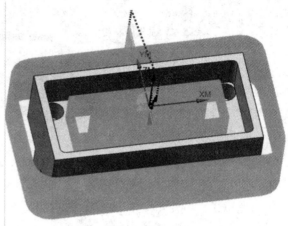

Figure 1-38　Tool Path Generation Diagram

1.4.2　Programming for reroughing

Since the radius of the concave fillet in the cavity is 4 mm, a flat-end tool with a diameter of 6 mm is selected for the secondary roughing of fillet residue. The steps are as follows:

(1) open the [Machine Tool] view, and then follow the previous step to create a carbide tool with a diameter of 6 mm, tool number 2, and name the tool D6.

(2) The name used by the second roughing program in NX 12.0 software is [Rest Milling], as shown in Figure 1-39. The function of rest milling is to remove the remaining part after the previous operation or machining by the last tool. [Rest Milling] is a branch of [Cavity Mill]. We can obtain the machining method of [Rest Milling] simply by modifying several parameters in the program of [Cavity Mill]. In practical application, the method editing program that creates [Rest Milling] directly from [Create Operation] is generally not used, because the program opened in this way needs to reset all the options in the initialization interface. The steps are so tedious that the programming may be delayed. We use the method of copying the previous program, and then modify the parameters of the program to obtain the machining method of [Rest Milling]. In this way, the parameters of the original program can be copied together, and it is only necessary to modify the parameters with changes, thus it can reduce the workload of programming and improving the programming efficiency.

(3) In the [Machine Tool] view, right-click the [CAVITY_MILL] program that has just been completed, and then right-click [Copy], as shown in Figure 1-40, circle 2. Next, right-click [D6] tool and select [Paste Inside] as shown in Figure 1-41, circle 4. A new [Cavity Mill] program that prompts error, as shown in Figure 1-42, circle 5.

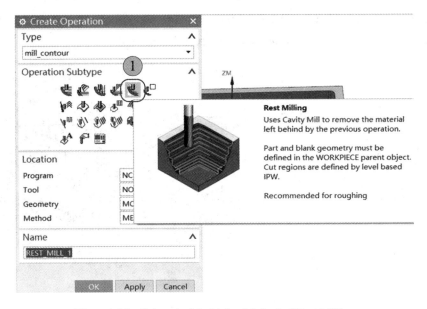

Figure 1-39　Select the Machining Method of Rest Milling

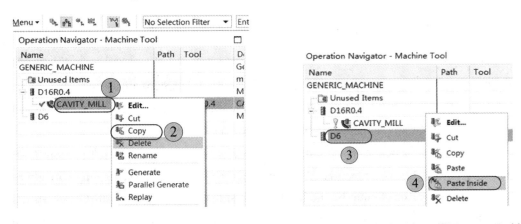

Figure 1-40　Copy the [Cavity Mill] Program　　Figure 1-41　Paste the [Cavity Mill] Program Inside

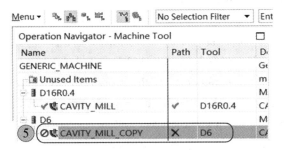

Figure 1-42　New [Cavity Mill] Program

(4) Double-click to open the newly copied [Cavity Mill] program. Set the [Maximum Distance] at the level in Z direction to "0.2 mm", as shown in Figure 1-43.

(5) Click [Cutting Parameters] to modify the settings, as shown in Figure 1-44.

Figure 1-43 Set the [Maximum Distance] at the Level in Z Direction to "0.2 mm" per Cut

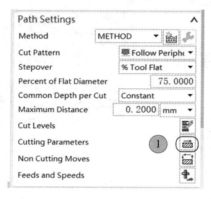

Figure 1-44　Click [Cutting Parameters]

(6) Click the [Containment] label in the [Cutting Parameters] dialog box to set the remaining blank of the workpiece. In the [In Process Workpiece] pull-down menu, select [Use Level Based] to set the blank to be machined as the remainder of the previous machining, which defaults to the [Use Level Based] option in [Rest Milling]. This is the only option that makes the difference between [Cavity Mill] and [Rest Milling]. Next, click [OK], as shown in Figure 1-45.

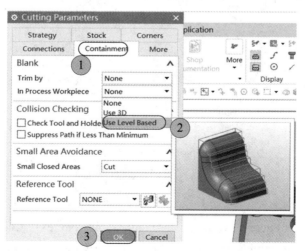

Figure 1-45　Settings of [Containment] Label

(7) Click [Cutting Levels] icon to set the machining depth. In the [Cutting Levels] dialog box, click ⊠ [Remove] icon on the right of the list to remove the original machining depth. Then click [Select Object] in [Range Definition] to select the floor of the workpiece cavity as the deepest position for machining (note: if this option is not selected, the intermediate round hole will be machined) as shown in Figure 1-46. Click [OK] to exit the [Cutting Levels] dialog box.

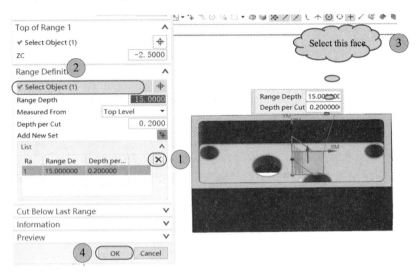

Figure 1-46　Set the Machining Depth

(8) Click [Feeds and Speeds] icon, set the [spindle Speed] to "4000 rpm", click [Calculator] icon on the right of the speed, and finally click [OK] button to exit the [Feeds and Speeds] dialog box, as shown in Figure 1-47.

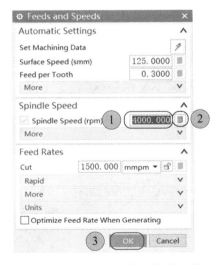

Figure 1-47　Set the [Spindle Speed]

(9) Click [Generate] icon, as shown in Figure 1-48, circle 1. Calculate the machining program, generate the machining path of [Rest Milling], and click [OK] to exit [Rest Milling], as shown in Figure 1-48, circle 2.

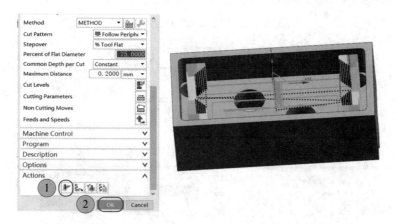

Figure 1-48　Generate the Tool Path of [Rest Milling]

1.4.3　Programming for floor finishing

The programming method for cavity floor finishing with a ϕ 10 carbide tool is as follows:

(1) Click to open the [Machine Tool] view, then follow the previous step to create a carbide tool with a diameter of 10 mm, tool number 3, and name it D10.

(2) Create a program for floor finishing, click 🛠 [Create Operation] icon, and select [mill_planar] in the [Type] pull-down menu and 凵 [Floor Wall Mill] in the [Operation Subtype] option. After that, select the [Bottom] program group in the [Program] option, select [D10] milling tool just created in the [Tool] option, and select [WORKPIECE] in the [Geometry] option. Finally, click [OK] to enter the [Floor Wall Mill] dialog box, as shown in Figure 1-49.

(3) Click 📦 [Specify Cut Area Floor] icon to select the floor to be machined. Click on the cavity floor, as shown in Figure 1-50, circle 2 and click [OK] to exit the [Cut Area] dialog box, as shown in Figure 1-50, circle 3.

(4) Check in front of the [Automatic Walls] to automatically capture the side wall of the workpiece adjacent to the selected floor as the wall geometry.

Figure 1-49　Enter the Step of [Floor Wall Mill]

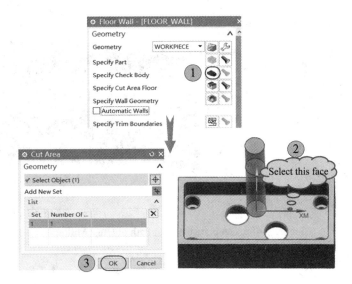

Figure 1-50　Select Floor in the [Cut Area] Dialog Box

(5) Since the machined surface is a concave plane, it is more suitable to use the method of [Follow Periphery], so the [Cut Pattern] is set to [Following Periphery], as shown in Figure 1-51, circle 1. The stepover in XY directions during finishing shall be less than 50% of the tool diameter. It defaults to 50% in the software, so no modification is required. [Floor Blank Thickness] is set to the stock of the floor during roughing, and is taken as 0.2 mm in this option, as shown in Figure 1-51, circle 2.

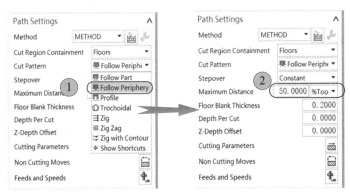

Figure 1-51　Selection of [Path Setting] Dialog Box

(6) Click ▦ [Cutting Parameters] icon to open the [Cutting Parameters] dialog box, and click the [Stock] label above and set the [Wall Stock] to "0.1 mm" (note: since this program is a floor finishing program, the [Final Floor Stock] should be "0 mm"). The finishing of the workpiece wall shall be carried out by a special program, so 0.1 mm is entered in the [Wall Stock] option, leaving a stock of 0.1 mm on one side for the program of wall finishing. To reduce the tolerance and improve the machining accuracy, set the [Internal Tolerance] and [External Tolerance] to "0.01 mm" respectively, and click [OK] to exit the [Cutting Parameters] dialog box, as shown in Figure 1-52.

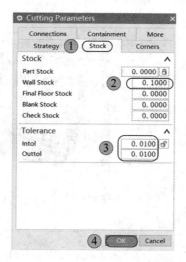

Figure 1-52　Settings of Machining [Stock]

(7) In the [Non Cutting Moves] dialog box, click the [Engage] label above to set engage parameters, as shown in Figure 1-53.

(8) Click 🐝 [Feeds and Speeds] icon to set the [Spindle Speed] to "3500 rpm", and then click the ▣ [Calculator] icon on the right. Set the [Feed Rates] to "1000 mmpm". Set the [Engage] option in [More] to "70%" to reduce the feed speed, as shown in Figure 1-54, circle 3, and then click [OK] to exit the [Feeds and Speeds] dialog box, as shown in Figure 1-54.

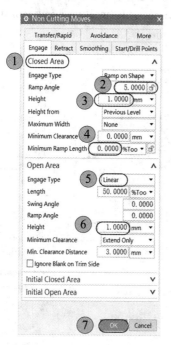

Figure 1-53　Settings of Engage Parameters

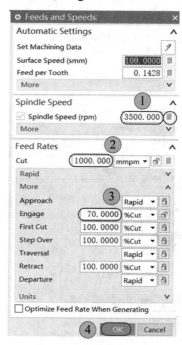

Figure 1-54　[Feeds and Speeds] Dialog Box

(9) Click ⊩ [Generate] icon to calculate the machining program, as shown in Figure 1-55, circle 1. Generate the machining path of [Floor Wall Mill], and click [OK] to exit [Floor Wall

Mill] setting, as shown in Figure 1-55, circle 2.

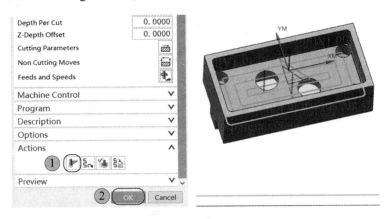

Figure 1-55 Generate the Machining Path of Floor Wall Mill

1.4.4 Programming for side finishing

The programming method for finishing the internal and external contours of parts with $\phi 6$ and $\phi 10$ carbide tools is as follow:

(1) To modify the floor finishing program to the side wall finishing program, first right-click the completed [Floor Wall Mill] program as shown in Figure 1-58, circle 2 in the [Machine Tool] View. Select the [Copy] option as shown in Figure 1-56, circle 3, then right-click [D6] to select [Paste Inside Paste] as shown in Figure 1-57, circles 4 and 5, and finally a newly formed [Floor Wall Mill] program that prompts the error, as shown in Figure 1-58, circle 6.

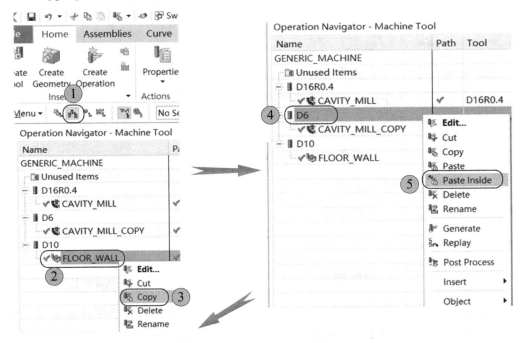

Figure 1-56 Copy the [Floor Wall Mill] Program   Figure 1-57 Paste the [Floor Wall Mill] Program Inside

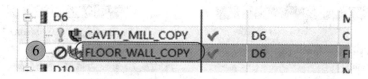

Figure 1-58 Copy the New [Floor Wall Mill] Program

(2) Double-click to open the newly copied [Floor Wall Mill] program. Set the [Cut Pattern] to [Profile], as shown in Figure 1-59.

(3) Click  [Cutting Parameters] icon to set the [Wall Stock] in the [Stock] label to "0 mm". Click [OK] to exit the [Cutting Parameters] dialog box, as shown in Figure 1-60.

Figure 1-59 Select [Profile] Machining Method   Figure 1-60 Set the [Wall Stock] to "0"

(4) Click  [Non Cutting Moves] icon, select the [Engage] label, and set the [Engage Type] in closed area to [Same As Open Area] and that in [Open Area] to [Arc]. The [Radius] is set to "50%" of the tool diameter, the [Arc Angle] set to "90 degrees", the [Height] set to "1 mm", and the [Minimum Clearance] set to [None]. As the tool radius compensation function needs to be enabled in finishing, G41 and G42 tool compensation must be established on straight line statements, so the option [Start at Arc Center] for arc engage type is selected to establish the straight extension line of the engage arc. If this item is not opened, there will be no straight line statement when setting the tool radius compensation, so the tool compensation will be established on the arc, and an alarm will be generated during the machining of the machine tool. The specific parameter settings are shown in Figure 1-61.

(5) Click the [More] label as shown in Figure 1-62, circle 5 to open the tool radius compensation function, set the [Cutcom Location] to [All Finish Passes], cancel the values in the options of [Minimum Move] and [Minimum Angle] and rewrite them to "0 mm", and then click [OK] to exit the [Non Cutting Moves] dialog box.

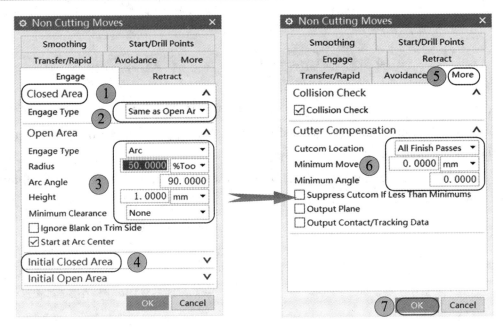

Figure 1-61   Setting of Engage Arc   Figure 1-62   Open the Tool Radius Compensation Function

(6) Click the 🐾 [Feeds and Speeds] icon to pop up [Feeds and Speeds] dialog box and modify the [Feed Rates] [Cut] to "500 mmpm" (the feeding speed should not be too fast during side wall finishing, and shall be controlled below 500 mmpm; otherwise the machining surface will be too rough and affect the machining quality). Next, click [OK] to exit the [Feeds and Speeds] dialog box, as shown in Figure 1-63.

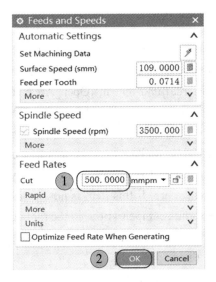

Figure 1-63   Set the [Feed Rates]

(7) Click the 📭 [Generate] icon as shown in Figure 1-64, circle 1, calculate the machining program and generate the program of [Side Wall Finishing] as shown in Figure 1-64, circle 2, and then click [OK] to exit [Floor Wall Mill] settings.

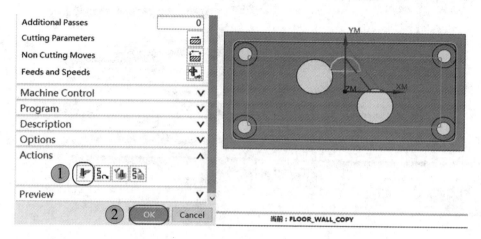

Figure 1-64　Generate the Program of [Side Wall Finishing]

(8) Finish the profile with a φ10 carbide tool. Click [Create Operation] icon, select [mill_planar] in the [Type] pull-down menu and 🔧 [Finish Walls] in the [Operation Subtype] option. After that, select [BOTTOM] in the [Program] option, select [D10] in the [Tool] option, and select [WORKPIECE] in the [Geometry] option. Finally, click [OK] to enter the [Finish Walls] dialog box, as shown in Figure 1-65.

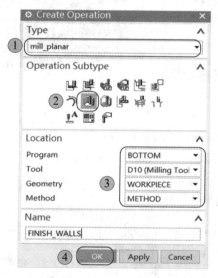

Figure 1-65　Create the Program of [Finish Walls]

(9) Click 📦 [Specify Part Boundaries] icon to open the [Part Boundaries] dialog box, as shown in Figure 1-66, circle 1. In the [Boundaries Selection Method] option, select [Curves] to create machining boundaries, as shown in Figure 1-66, circle 2. Set the [Boundary Type] to [Closed], the [Tool Side] to [Outside], and the [Plane] to [Automatic], as shown in Figure 1-66, circle 3. Next, manually select the outer circle line of contour as the machining boundary, and click [OK] to exit the setting, as shown in Figure 1-66, circles 4, 5 and 6.

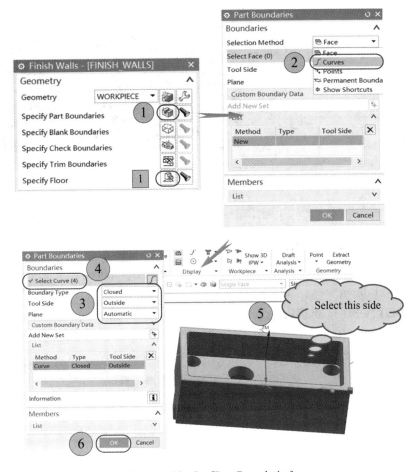

Figure 1-66　Set [Part Boundaries]

(10) Click 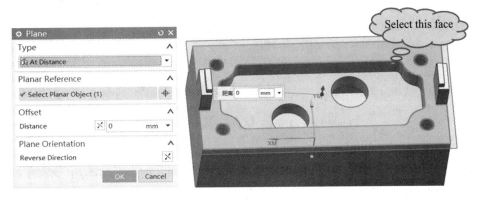 [Specify Floor] icon shown in Figure 1-66, box 1 to open the [Plane] dialog box. Select the floor plane as the machining floor, as shown in Figure 1-67.

Figure 1-67　Select the Machining Floor

(11) Click [Cutting Levels] icon to open the [Cutting Levels] dialog box and select [Constant] in the [Type] pull-down menu. The height of each level is fixed, and the [Depth Per Cut] [Common] is set to "14 mm." After that, click [OK] to exit the setting, as shown in Figure 1-68.

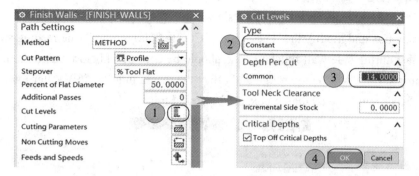

Figure 1-68　Set the [Depth Per Cut] to "14 mm"

(12) Click 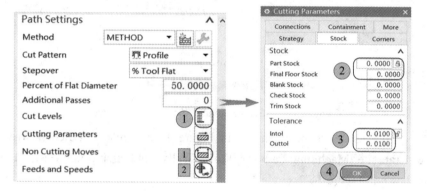 [Cutting Parameters] icon, select the [Stock] option in the pop-up [Cutting Parameters] dialog box, set the [Part Stock] and [Final Floor Stock] to "0", set the [Tolerance] [Intol] and [Outtol] to "0.01", and click [OK] to exit the settings, as shown in Figure 1-69.

Figure 1-69　Settings of Machining Stock and Internal and External Tolerance

(13) Click 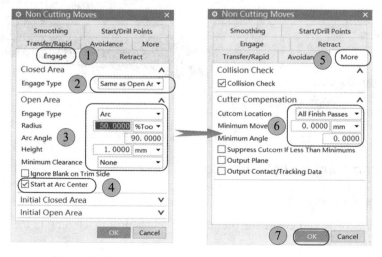 [Non Cutting Moves] icon as shown in Figure 1-69, box 1 to open [Non Cutting Moves] dialog box, set the [Engage Type] to [Arc] and enable the radius compensation function. The specific setting method is the same as [Floor Wall Mill], as shown in Figure 1-70.

Figure 1-70　Setting of [Non Cutting Moves] Dialog Box

(14) Click 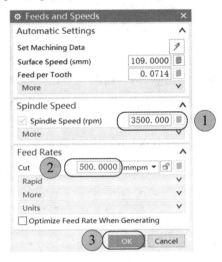 [Feeds and Speeds] icon as shown in Figure 1-69, box 2 to open the [Feeds and Speeds] dialog box and modify the [Feed Rates] [Cut] to "500 mmpm" (the feeding speed should not be too fast during side wall finishing, and should be controlled below 500 mmpm; otherwise the machining surface will be too rough and affect the machining quality). After that, click [OK] to exit the [Feeds and Speeds] dialog box, as shown in Figure 1-71.

Figure 1-71　Set the Spindle Speed and Feed Rate

(15) Click 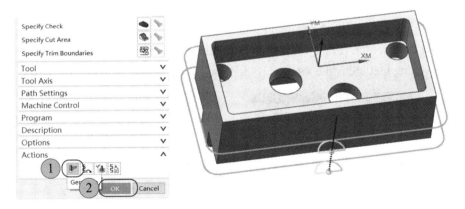 [Generate] icon, as shown in Figure 1-72, circle 1. Calculate the machining program, generate the [Machining Path] of [Finish Walls], and click [OK] to exit [Finish Walls] settings, as shown in Figure 1-72, circle 2.

Figure 1-72　Generate the Program of [Finish Walls]

1.4.5　Programming for hole making

1. Programming for center drill

Create a $\phi 6$ center drill, and create the program for center drill machining in the hole position of the workpiece.

(1) To create a center drill, click [Machine Tool] icon to switch the [Operation Navigator] to the [Machine Tool] view. Click [Create Tool] icon to open the [Create Tool] dialog box, as shown in Figure 1-73, circles 1 and 2.

Figure 1-73   [Create Tool] Dialog Box

(2) In the [Type] pull-down menu, select [hole_making]. Select [Center Drill] icon in the [Tool Subtype] dialog box, enter the tool name [Center Drill D6] in the [Name] dialog box, and then click [OK] as shown in Figure 1-73, circles 3 and 4 to open the [Center Drill Tool] dialog box.

(3) Set the tool [Diameter] to "6 mm", set the [Tool Number] and [Adjust Register] number to "4", and set other parameters by default. Finally, click [OK] to exit the settings, as shown in Figure 1-74.

(4) To create a program for center drill, click [Create Operation] icon to open [Create Operation] dialog box, and select [hole_making] in the [Type] pull-down menu and [Center Drill] in the [Operation Subtype] option. After that, select [Bottom] in the [Program] option, select [Center Drill D6] just created in the [Tool] option, and select [WORKPIECE] in the [Geometry] option. Finally, click [OK] to enter the [Center Drill] dialog box, as shown in Figure 1-75.

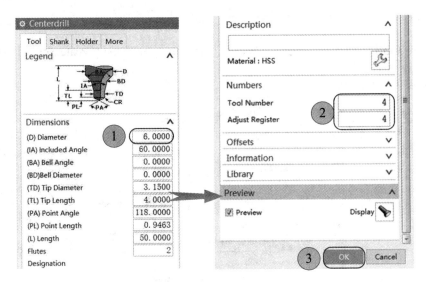

Figure 1-74 Setting of [Centerdrill]

(5) In the [Centerdrill] dialog box, click 👆 [Specify Feature Geometry] icon, as shown in Figure 1-76, circle 5. Select the six hole positions to be subject to center drilling, and press and hold the [Shift] key on the keyboard. Click the bottom hole and the top hole in the list and release the [Shift] button. At this point, all the holes in the list appear in a blue background. Click the 🔒 icon and select the 🔓 User Defined option, set the [Depth] to "1 mm" as the hole depth, and click [OK] to exit the settings, as shown in Figure 1-77.

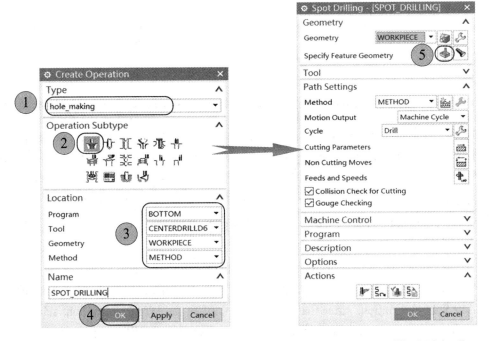

Figure 1-75 Select the Program of Center Drill   Figure 1-76 [Spot Drilling] Dialog Box

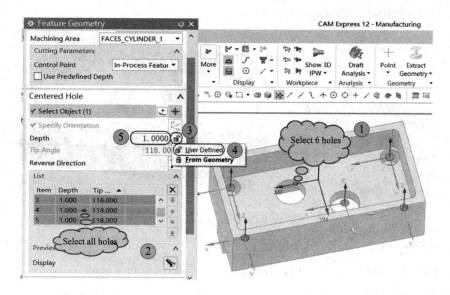

Figure 1-77   Select Holes to Set the Machining Depth

(6) Click [Cutting Parameters] icon to open [Cutting Parameters] dialog box, and set the [Top Offset] [Distance] to "1 mm" so as to reduce the initial distance of drilling and increase the machining speed. Next, click [OK] to exit the dialog box, as shown in Figure 1-78.

(7) Click [Feeds and Speeds] icon to open [Feeds and Speeds] dialog box, set the [Spindle Speed] to "1200 rpm", click the [Calculator] icon on the right of the speed and set the [Feed Rate] [Cut] to "100 mmpm", and finally click [OK] button to exit the dialog box, as shown in Figure 1-79.

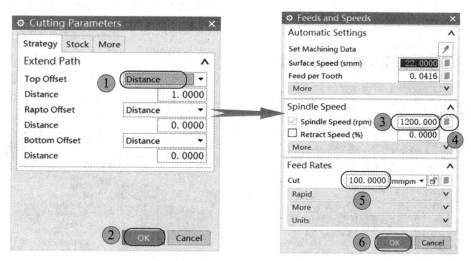

Figure 1-78   Set the Initial Distance of Drilling   Figure 1-79   Set the Spindle Speed and Feed Rate

(8) Click [Generate] icon as shown in Figure 1-80, circle 1, calculate the machining program and generate the path of [Spot Drill], and then click [OK] to exit [Spot Drill] settings, as shown in Figure 1-80.

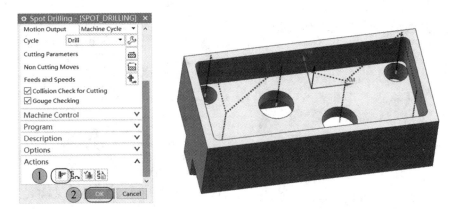

Figure 1-80　Generate the Program of Spot Drill Machining

2. Programming for drilling of φ4.5 circular holes

Create a φ4.5 drill and edit the drilling program of 4 φ4.5 circular holes.

(1) In the [Machine Tool] view, create a φ4.5 drill using the method of creating the center drill. Set the [Tool Number] to "5" and the [Tool Name] to [Drill D4.5], as shown in Figure 1-73 and Figure 1-74.

(2) Create a drill program, click [Create Operation] icon to open [Create Operation] dialog box, select [hole_making] in the [Type] pull-down menu and ⊕ [Drilling] in the [Operation Subtype] option, as shown in Figure 1-81, circle 1. Select [Bottom] in the [Program] option, select [Drill D4.5] just created in the [Tool] option, and select [WORKPIECE] in the [Geometry] option. Finally, click [OK] as shown in Figure 1-81, circle 3 to enter the [Drilling] dialog box as shown in Figure 1-82.

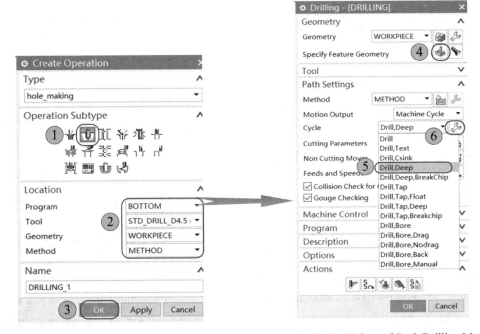

Figure 1-81　Create a Drill Program　　　　　Figure 1-82　Select Holes and Peck Drilling Method

(3) In the [Drilling] dialog box, click [Specify Feature Geometry] icon, as shown in Figure 1-82, circle 4. In the pop-up dialog box, select four φ4.5 holes, and then press and hold the [Shift] key on the keyboard. Click the bottom hole and the top hole in the list and release the [Shift] button. At this point, all the holes in the list appear in a blue background. Click 🔒 icon to select the ⚬ User Defined option, set the [Depth] to "17 mm" as the drilling [Depth], and click [OK] to exit the settings, as shown in Figure 1-83.

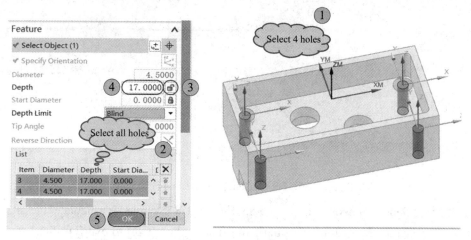

Figure 1-83　Select Four Holes to Set the Machining [Depth] to "17 mm"

(4) Click 📄 [Cutting Parameters] icon, set the [Top Offset] [Distance] to "1 mm" so as to reduce the initial distance of drilling, and click [OK] to exit the dialog box, as shown in Figure 1-84.

(5) Select [Drill, Deep] in the [Cycle] pull-down menu and set it to peck drill, as shown in Figure 1-82, circle 5. Click 🔧 [Edit Cycle] icon as shown in Figure 1-82, circle 6, and set the [Step] [Maximum Distance] to "1 mm" as shown in Figure 1-85, circle 3. This value is the depth of each layer in peck drilling. Finally, click [OK] to exit the dialog box, as shown in Figure 1-85.

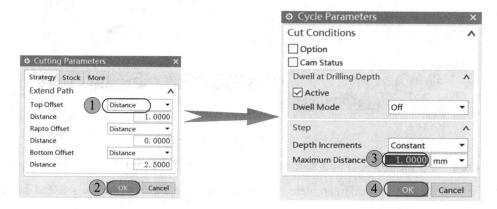

Figure 1-84　Set the Initial Distance of Drilling　　　Figure 1-85　Set the Depth of Each Layer of Peck Drilling

(6) Click 🔧 [Feeds and Speeds] icon, set the [Spindle Speed] to "1000 rpm", then click the [Calculator] icon on the right of the speed and set the [Feed Rates] [Cut] to "100 mmpm", and

finally click [OK] button to exit the dialog box, as shown in Figure 1-86.

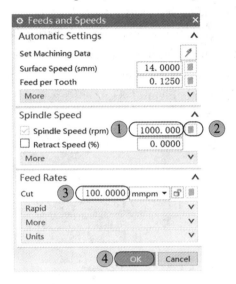

Figure 1-86   Set the Spindle Speed and Feed Rate

(7) Click [Generate] icon as shown in Figure 1-87, circle 1, calculate the machining program and generate the path of [Drilling], and finally click [OK] to exit [Drilling] settings, as shown in Figure 1-87, circle 2.

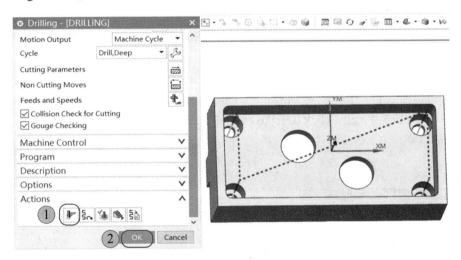

Figure 1-87   Generate the Program of Drilling

3. Programming for four $\phi 8$ counterbores with a depth of 10 mm

Create a $\phi 8$ flat-end tool and complete the machining program of four $\phi 8$ flat-end holes on workpiece.

(1) In the [Machine Tool] view, create a $\phi 8$ flat-end tool using the method of creating the flat-end tool. Set the tool number to 6 and the tool name to [D8], as shown in Figure 1-17 to Figure 1-20.

(2) To modify the Φ4.5 drill program to Φ8 counterbore program, first right-click the [Drilling] program just-completed in the [Machine Tool] view, and select [Copy], as shown in Figure 1-88. Next, right-click [D8] and select [Paste Inside], as shown in Figure 1-89, circle 3 and 4. Then a new [Drilling] program prompts error, as shown in Figure 1-90.

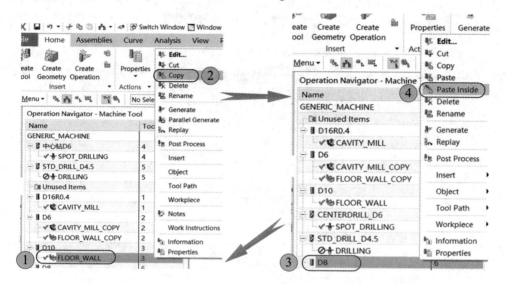

Figure 1-88　Copy the [Drilling] Program　　　Figure 1-89　Paste the [Drilling] Program Inside

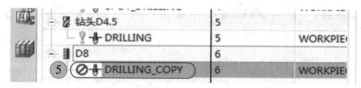

Figure 1-90　Copy the New [Drilling] Program

(3) Double-click to open the newly copied drilling program. In the [Drilling] dialog box, click 🔧 [Specify Feature Geometry] icon to modify the depth to the counterbore [Depth] to "10 mm". Click [OK] to exit the setting, as shown in Figure 1-91.

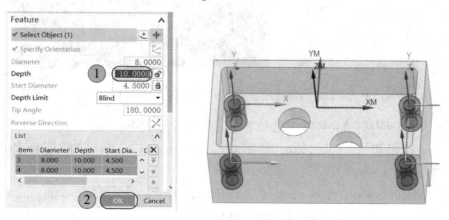

Figure 1-91　Modify the Drilling [Depth] to "10 mm"

(4) Click ![icon] [Feeds and Speeds] icon to open [Feeds and Speeds] dialog box, set the [Spindle Speed] to "2000 rpm", then click the [Calculator] icon on the right of the speed and set the [Feed Rates] [Cut] to "60 mmpm", and finally click [OK] button to exit the dialog box, as shown in Figure 1-92.

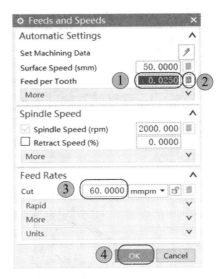

Figure 1-92 Set the Spindle Speed and Feed Rate

(5) Click ![icon] [Generate] icon to calculate the machining program, generate the path of [Drilling], and click [OK] to exit [Drilling] settings, as shown in Figure 1-93.

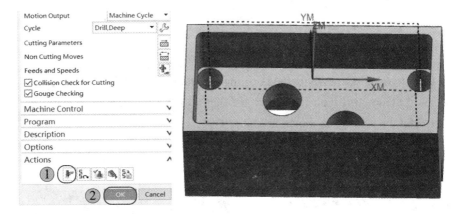

Figure 1-93 Generate the Program of Counterbore Machining

4. Create the machining program of two φ12 through-holes

Create a φ12 drill and complete the machining program of two φ12 through-holes on workpiece.

(1) In the [Machine Tool] View, create a φ13 drill using the method of creating a drill. Set the [Tool Number] to "7" and the [Tool Name] to [Drill D13], as shown in Figure 1-73 and Figure 1-74.

(2) To modify the Φ4.5 drill program to Φ13 through-hole program, first click the [Machine Tool] view as shown in Figure 1-94, circle 1, then right-click the [Drilling] program of D4.5 drill as shown in Figure 1-94, circle 2, and select [Copy] as shown in Figure 1-94, circle 3. Next, right-click [Drill D13] and select [Paste Inside], as shown in Figure 1-94, circles 4 and 5, and a new [Drilling] program prompts error, as shown in Figure 1-94, circle 6.

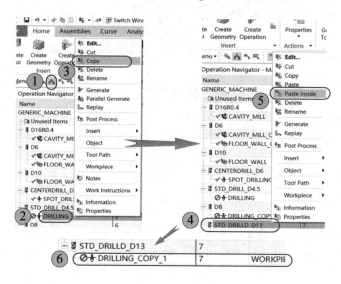

Figure 1-94　Copy the Path of New [Drilling] Tool

(3) Double-click to open the newly copied drilling program. In the [Drilling] dialog box, click 🔧 [Specify Feature Geometry], click ✖ icon in the list to delete all the holes previously selected, and re-select two Φ13 holes. Define and modify the depth to the drilling [Depth] to "12 mm", and click [OK] to exit the setting, as shown in Figure 1-95.

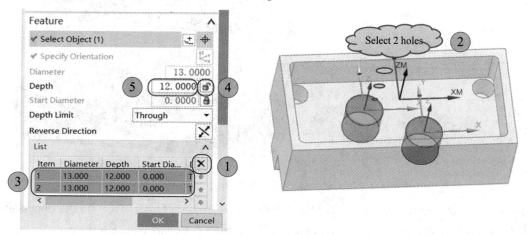

Figure 1-95　Select Two Φ13 Holes and Set the Hole [Depth] to "12 mm"

(4) Click 🔧 [Feeds and Speeds] icon, set the [Spindle Speed] to "500 rpm", then click the [Calculator] icon on the right of the speed and set the [Feed Rates] [Cut] to "100 mmpm", and finally click [OK] button to exit the dialog box, as shown in Figure 1-96.

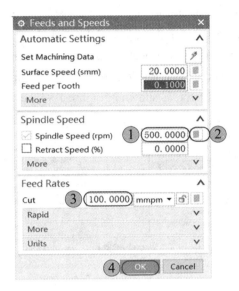

Figure 1-96　Set the Spindle Speed and Feed Rate

(5) Click ⚐ [Generate] icon as shown in Figure 1-97, circle 1, calculate the machining program and generate the path of [Drilling], and finally click [OK] to exit [Drilling] settings, as shown in Figure 1-97, circle 2.

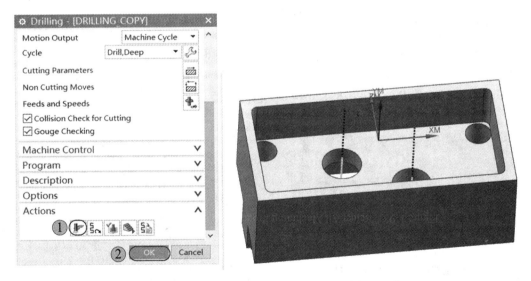

Figure 1-97　Generate the [Drilling] Program of φ13 Holes

(6) Simulate the edited machining program, click ▦ [Program Order] view on the upper left, and select all the edited machining programs in the [Program Order] view interface, as shown in Figure 1-98, circle 2.

(7) Click ▨ Verify Tool Path icon on the [Home] as shown in Figure 1-99, circle 1 and 2 to open the [Tool Path Visualization] dialog box, as shown in Figure 1-100.

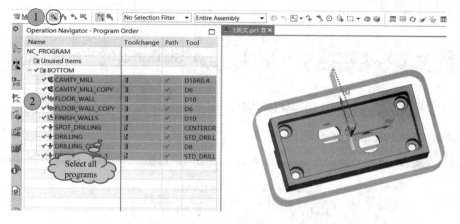

Figure 1-98　Select All Machining Programs in the [Program Order] View

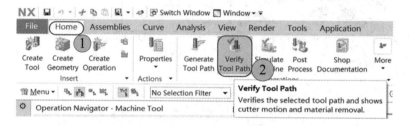

Figure 1-99　Click [Verify Tool Path] Icon

(8) Click [3D Dynamic] to switch the animation to 3D model, and then click ▶ [Play] icon　below
to complete the machining simulation of workpiece floor path, as shown in Figure 1-100, circle 2.

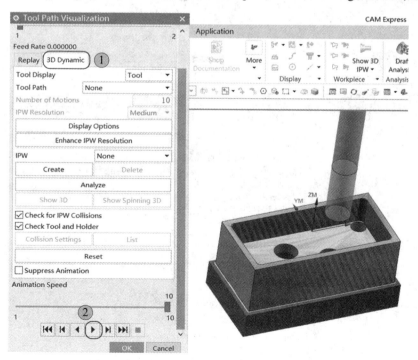

Figure 1-100　Simulate the Machining Path of Floor

1.5　Step 2 of machining of switch box frame

1.5.1　Creation of parent node group

1. Creation of program group

First click [Program Order] view icon as shown in Figure 1-101, circle 1 to switch the

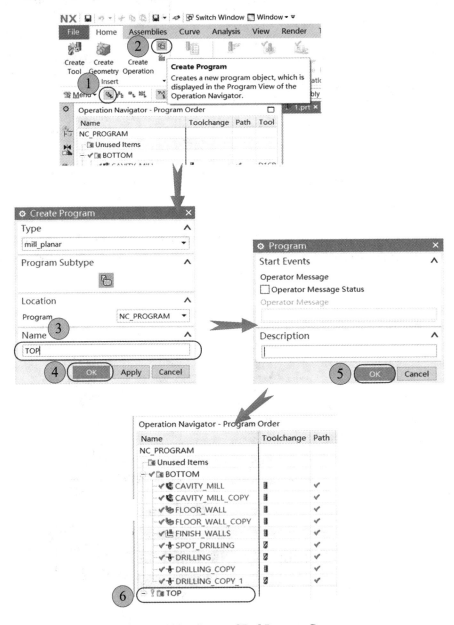

Figure 1-101　Create a [Top] Program Group

[Navigator] to [Program Order] view, then click 🖼 [Create Program] icon on the upper left as shown in Figure 1-101, circle 2, and enter the name "Top" in the pop-up dialog box as shown in Figure 1-101, circle 3. Click [OK] to open the [Program] dialog box, as shown in Figure 1-101, circle 4, and then click [OK] to exit the [Program] dialog box, as shown in Figure 1-101, circle 5. Add a program group [Top] under the navigator of [program order] view, as shown in Figure 1-101, circle 6.

2. Creation of coordinate system and geometry

The steps for creating the coordinate system and geometry of Step 2 are as follows:

(1) Switch the [Operation Navigator] to [Geometry] view, right-click [MCS_MIL] coordinate system and select [Copy], and then right-click [GEOMETRY] and select [Paste Inside]. At this point, make a copy of the previous coordinate system, the geometry and the program, as shown in Figure 1-102.

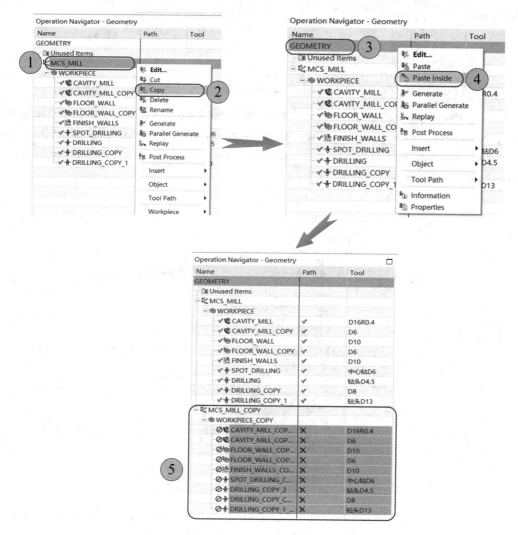

Figure 1-102　Copy the Coordinate System, Geometry and Program of Step 1 Operation

(2) To modify the coordinate system and the geometry name, right-click [MCS_MILL], select [Rename] and modify the coordinate system name to [Bottom MCS], and then right-click [WORKPIECE], select [Rename] and modify the geometry name to [Bottom], as shown in Figure 1-103.

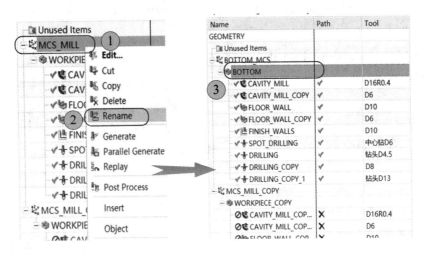

Figure 1-103　Copy the Coordinate System and Geometry Names of Step 1 Operation

(3) Modify the coordinate system and geometry names of Step 2 operation to [TOP_MCS] and [Top] respectively in the same way. Press and hold the [Ctrl] key to select all programs in the [Top] geometry that prompt errors, and right-click [Delete] to delete all copied programs, as shown in Figure 1-104.

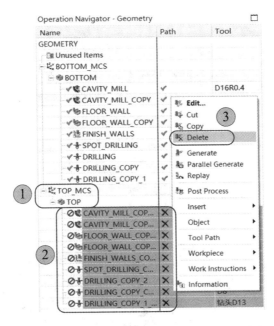

Figure 1-104　Modify the Coordinate System and Geometry names of Step 2 Operation and Delete Copied Programs

(4) Modify the Z-axis direction of the coordinate system, double-click ⊟ [TOP_MCS] icon and select [Coordinate System] icon in the pop-up [MCS Mill] dialog box, as shown in Figure 1-105.

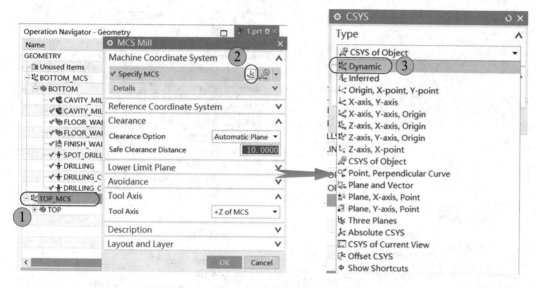

Figure 1-105   Set the Coordinate System of Step 2 Operation   Figure 1-106   Select the Dynamic Mode

(5) Select icon   in the [Coordinate System] dialog box as shown in Figure 1-106, circle 3, and double-click the arrow on the ZM of the coordinate system to keep the coordinate direction upward, as shown in Figure 1-107, circle 1.

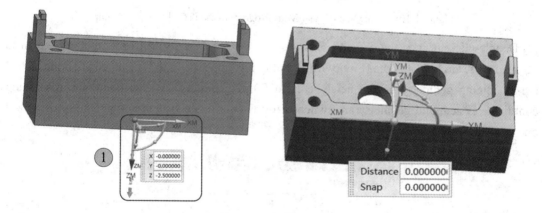

Figure 1-107   Create the Coordinate System of Step 2 Operation by Double-clicking the ZM Arrow to Reverse It

1.5.2   Programming for roughing

The steps for roughing the top surface of parts using $\phi 16R0.4$ drilling and milling tool are as follows:

(1) Edit the roughing program of Step 2 operation in the [program Order] view. First, copy the

roughing program [CAVITY_ALL] under the [Bottom] program group as shown in Figure 1-108, circle 1 and circle 2, and then right-click [Top] and select [Paste Inside] as shown in Figure 1-108, circle 3 and 4, and finally form a new roughing program as shown in Figure 1-108, circle 5.

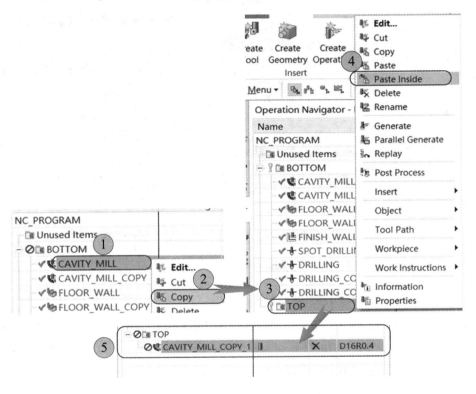

Figure 1-108　Copy the [Cavity Mill] Program to the [TOP] Program Group

(2) Double-click to open the copied ⊘⚙ CAVITY_MILL_COPY_1 [Cavity Mill] program, as shown in Figure 1-108, circle 5, and the [Cavity Mill] dialog box appears. In the [Geometry] option, select the machining geometry as the newly created [Top] geometry. At this point, the machining coordinate system and geometry are switched to the coordinate system and geometry of Step 2 operation, as shown in Figure 1-109.

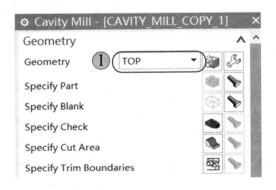

Figure 1-109　Select the Geometry of Step 2 Operation

(3) Click 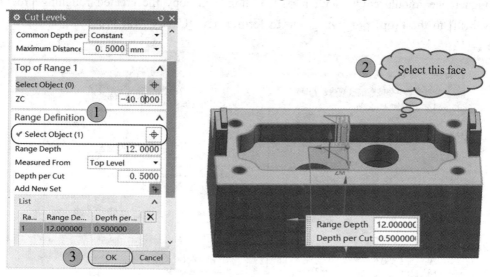 [Cutting Levels] icon and set the machining depth of Step 2 operation. Since most part of the floor has been machined in step 1 operation, it is only necessary to machine to the floors of two raised platforms for the outer frame. Open the [Cutting Levels] dialog box and directly click the plane as shown in Figure 1-110, circle 2, and the machining [Range Depth] is measured as "12 mm". Click [OK] to exit the [Cutting Levels] dialog box.

Figure 1-110   Select the Machining Floor

(4) The rest of the parameters do not need to be modified. Click 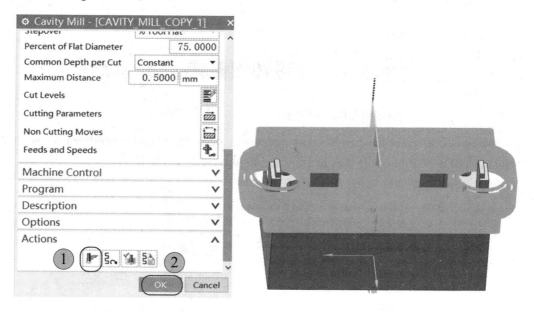 [Generate] icon to calculate the new machining path, as shown in Figure 1-111, circle 1. Click [OK] to exit the [Cavity Mill], as shown in Figure 1-111, circle 2.

Figure 1-111   Generate the Roughing Program of Step 2 Operation

1.5.3　Programming for reroughing

The machining method of rest milling is used to complete the programming of reroughing of workpiece.

(1) Roughen the middle cavity with a φ10 milling, and copy the created roughing program [Cavity Mill] to the [Top] program group to form a new [Cavity Mill Program], as shown in Figure 1-112.

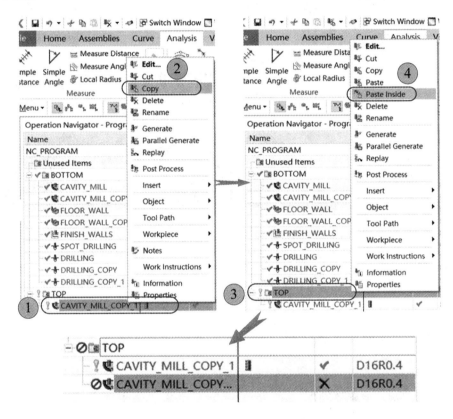

Figure 1-112　Copy the New [Cavity Mill] Program

(2) Double-click to open the newly copied program, click the [Tool] drop-down arrow, and change the tool to [D10] carbide tool, as shown in Figure 1-113, circle 1.

Figure 1-113　Modify the [Tool] to [D10 Milling Tool]

(3) Click [Cutting Levels] icon to set the initial and end depth of machining. First click the  [Select Object] icon in the [Top of Range 1] option and select the plane shown in Figure 1-114, circle 2 and 4. Then click [Select Object] icon in the [Range Definition] option and select the plane as shown in Figure 1-114, circle 4. The machining [Range Depth] is measured as "7 mm". Click [OK] to exit settings.

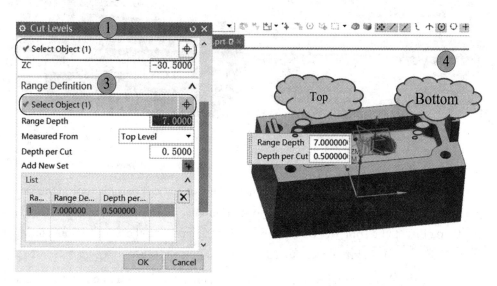

Figure 1-114   Select the Top and Bottom of Machining Range

(4) Click [Cutting Parameters] icon   to open the [Cutting Parameters] dialog box, click the [Containment] label and set the [Blank] [Trimby] as [Silhouette], so that the program can automatically trim the programs outside the workpiece range, as shown in Figure 1-115.

Figure 1-115   Set the [Trim by] to [Silhouette]

(5) Click ⊫ [Generate] icon to calculate the machining program, which generates the path of [Cavity Mill], and click [OK] to exit [Cavity Mill] settings, as shown in Figure 1-116.

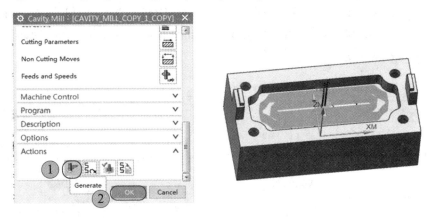

Figure 1-116 Generate the Roughing Path of φ10 Tool

(6) Prepare the program for reroughing of internal fillet using a φ4 carbide tool in the same way as step 5. First create a φ4 flat-end tool, with tool number 8. Copy the previous [Cavity Mill] program to modify the [Tool] to [D4 (Milling Tool)], and set the Z-direction cutting level [Maximum Distance] to "0.2 mm" per cut, as shown in Figure 1-117.

(7) Click the [Cutting Parameters] icon, select the [Containment] label in the [Cutting Parameters] dialog box, and set the [Reference Tool] to [D10 Milling Tool] as shown in Figure 1-118.

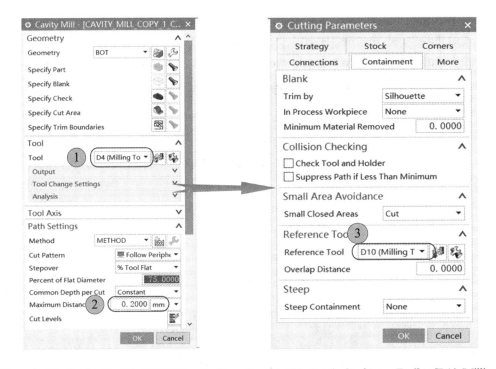

Figure 1-117 Set the [Tool] to [D4 (Milling Tool)]   Figure 1-118 Set the [Reference Tool] to [D10 (Milling Tool)]

(8) Set the [Spindle Speed] to "5000 rpm", click 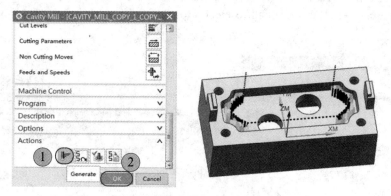 [Generate] icon to calculate the machining program and generate the machining path of [Cavity Mill], and click [OK] to exit [Cavity Mill] settings, as shown in Figure 1-119.

Figure 1-119　Program for Roughing of D4 Flat-end Tool

1.5.4　Programming for floor finishing

Complete the finishing of workpiece floor plane using a φ 10 carbide tool.

(1) Edit the program for floor finishing of Step 2 operation under the [Operation Navigator—Program Order] view. First right-click the floor finishing program [FLOOR_WALL] under the [Bottom] program group as shown in Figure 1-120, circles 1 and 2, then right-click to [Paste Inside] under the [Top] program group as shown in Figure 1-120, circles 3 and 4, and finally a new [Floor Wall Mill] program pops up as shown in Figure 1-120, circle 5.

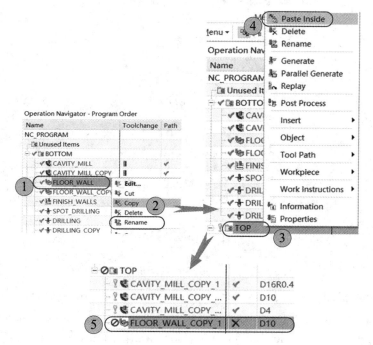

Figure 1-120　Copy the Floor Finishing Program of Step 1 Operation into the Program Group of Step 2 Operation

(2) Double-click to open the newly copied [Floor Wall Mill] program as shown in Figure 1-120, circle 5, and modify the parameters as follows: Click ⬡ [Specify Cut Area Floor] icon to open the [Cut Area] dialog box, click the icon ✗ to delete the machining floor selected last time as shown to Figure 1-121. circle1, and reselect a total of 6 machining floors, as shown in Figure 1-121, circle 3.

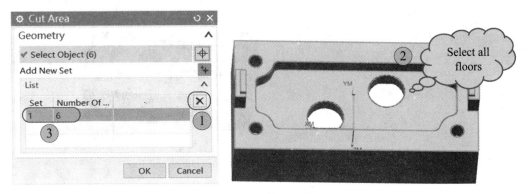

Figure 1-121　Reselect the Machining Floor

(3) Click ⬚ [Cutting Parameters] icon and set the [Tool Overhang] under the [Containment] label to "100%". If not set, an alarm will be raised when the program is generated. After setting the value to 100%, the tool machining range will be extended to all workpiece surfaces to avoid missing surfaces. The operation steps are shown in Figure 1-122.

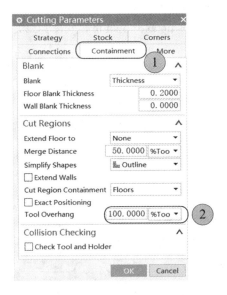

Figure 1-122　Set the [Tool Overhang] to "100%"

(4) The settings of all the other parameters will follow the original program without modification by default. Finally, click ⮕ [Generate] icon to calculate the machining program and generate the path of [Floor Wall], and click [OK] to exit [Floor Wall] settings, as shown in Figure 1-123.

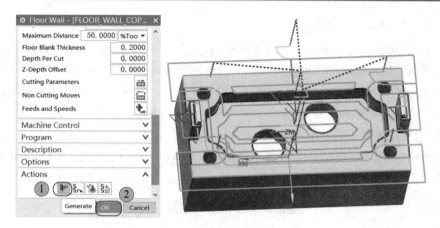

Figure 1-123　Generate the Program of Floor Finishing

1.5.5　Programming for side finishing

The method of editing the program for finishing all the side surfaces of Step 2 operation is as follows:

(1) Click the [Program Order] view to edit the program for side finishing of Step 2 operation. Copy the side wall finishing program [FLOOR_WALL_COPY] under the [Bottom] program group to the [Top] program group to form a new [Floor Wall Mill] program, as shown in Figure 1-124.

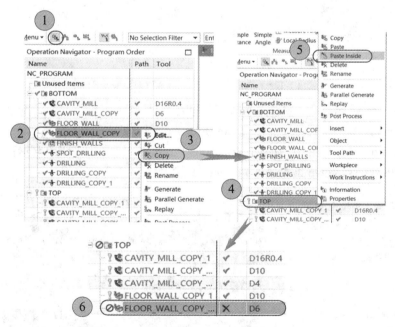

Figure 1-124　Copy the Side Finishing Program of Step 1 Operation into the Program Group of Step 2 Operation

(2) Double-click to open the newly copied bottom [Floor Wall Mill] program as shown in Figure 1-124, circle 6, and modify the parameters as follows: Click 🔲 [Specify Cut Area Floor] icon, then click ✖ to delete the previously machining floors selected, and reselect a total of 6

machining floors, as shown in Figure 1-121.

(3) Click 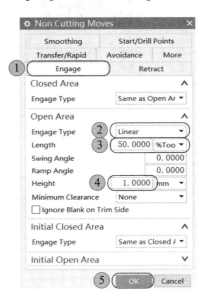 [Cutting Parameters] icon and set the [Tool Overhang] under the [Containment] label to "100%". If not set, an alarm will be raised when the program is generated. After setting the value to 100%, the tool machining range will be extended to all workpiece surfaces to avoid missing surfaces. The operation steps are shown in Figure 1-122.

(4) Set the [Engage Type] to [Linear Feed], click ☑ [Non Cutting Moves] icon and set the [Engage Type] under the [Engage] label to [Linear], the [Length] to "50%" of the tool diameter, and the [Height] to "1 mm". Click [OK] to exit the settings, as shown in Figure 1-125.

Figure 1-125　Set the [Engage Type] to [Linear Feed]

(5) The settings of all the other parameters will follow the original program without modification by default. Finally, click 🖙 [Generate] icon to calculate the machining program and generate the path of [Floor Wall Mill] as shown in Figure 1-126, circle 1, and click [OK] to exit [Floor Wall Mill] settings, as shown in Figure 1-126, circle 2.

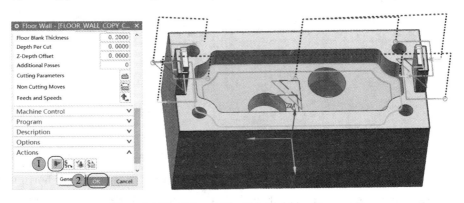

Figure 1-126　Generate the Side Finishing Program

1.5.6　Programming for angle-clearing of side finishing

Use a φ4 milling tool to edit the angle-clearing program for part side finishing.

(1) Edit the angle-clearing program for side finishing of Step 2 operation under the [Operation Navigator—Program Order] view. First right-click to copy the side wall finishing program [FLOOR_WALL_COPY_COPY] created in last step as shown in Figure 1-127, circles 2 and 3, then right-click to [Paste Inside] under the [Top] program group as shown in Figure 1-127, circles 4 and 5, and finally a new [Floor Wall Mill] program pops up as shown in Figure 1-127, circle 6.

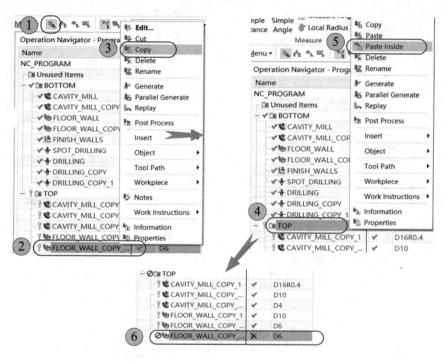

Figure 1-127　Copy the Side Finishing Program of Last Step into the [TOP] Program Group

(2) Double-click to open the newly copied [Bottom Wall Mill] program. Modify the parameters as follows: Click [Specify Cut Area Floor] icon to open the [Cut Area] dialog box, click to delete the machining floors of the previous program, and select new machining floors, as shown in Figure 1-128, circle 2.

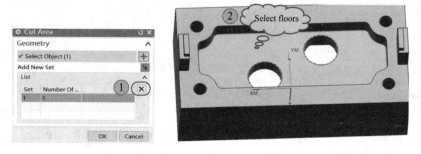

Figure 1-128　Select the Cavity Floor as Machining Floor

(3) Finish the cavity fillet with a φ4 tool, click to open the drop-down arrow on the right of the [Tool] option to expand the tool list, and select [D4 (Milling Tool)], as shown in Figure 1-129.

(4) The φ4 milling tool cannot be fed too deeply; otherwise it may be broken. Therefore, the [Depth Per Cut] in the tool [Path Setting] is set to "1 mm", and the angle-clearing part is machined in layers, as shown in Figure 1-130.

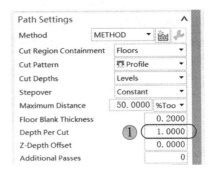

Figure 1-129　Select [D4 Milling Tool]　　　　Figure 1-130　Set [Depth Per Cut] to "1 mm"

(5) Click ⊞ [Cutting Parameters] icon　to open the [Cutting Parameters] dialog box, and set the [Blank] option in the [Containment] label to [3D IPW] (3D IPW is to use the remaining part after the preceding machining as the blank for this machining). Click [OK] to exit the settings, as shown in Figure 1-131.

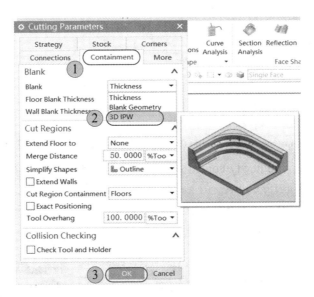

Figure 1-131　Set the Workpiece [Blank] to [3D IPW]

(6) The settings of all the other parameters will follow the original program without modification by default. Finally, click ⊮ [Generate] icon to calculate the machining program, and generate the path of [Floor Wall Mill], and click [OK] to exit [Floor Wall Mill] settings, as shown in Figure 1-132.

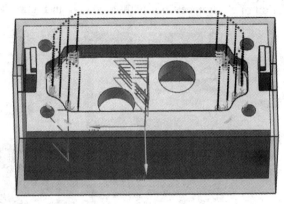

Figure 1-132　Generate the Angle-clearing Program of Finishing

(7) Simulate the edited machining program, click [Program Order] view at upper left, and select all the edited machining programs in the [Program Order] view interface, as shown in Figure 1-133.

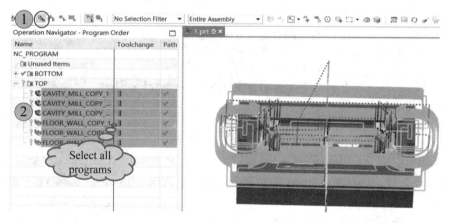

Figure 1-133　Select All Machining Programs in the [Program Order] View

(8) Click icon in the [Home] as shown in Figure 1-134 to open the [Tool Path Visualization] dialog box.

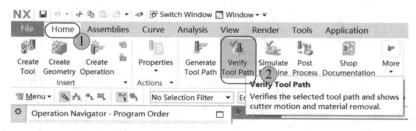

Figure 1-134　Click [Verify Tool Path] Icon

(9) In the pop-up [Tool Path Visualization] dialog box, click [3D Dynamic] to switch the animation to 3D model, and then click ▶ [Play] icon　below to complete the machining

simulation of workpiece floor path, as shown in Figure 1-135.

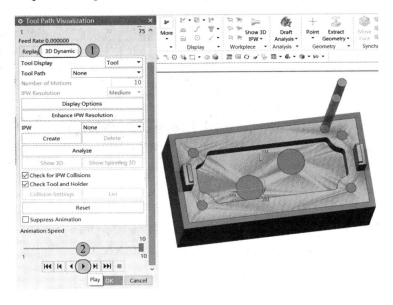

Figure 1-135　Simulation of the Machining Path of Floor Path

1.6　Generation of G code files

Generate all the edited machining programs into G code files.

(1) To install the NC postprocessor, click the [Menu] drop-down list to select [Tools] and scroll down the mouse wheel to select [Install NC Postprocessor]. The steps are as shown in Figure 1-136.

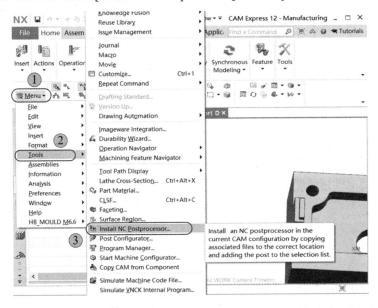

Figure 1-136　Install the NC Postprocessor

(2) In the [Select Postprocessor] dialog box that appears, select the location of the postprocessing file. Click the [Search Scope] pull-down menu and double-click to open the [FANUC 0i Postprocessing] folder in the disc directory, select the [FANUC0i.pui] postprocessing file in the folder, and click [OK] to exit the settings. The operation steps are shown in Figure 1-137.

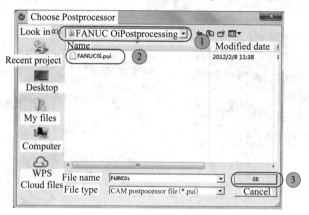

Figure 1-137　Select the Location of Postprocessing File

(3) In the pop-up [Install Postprocessor] dialog box, set the postprocessing name to "FANUC0i" and click [OK] to exit the settings. The operation steps are shown in Figure 1-138.

Figure 1-138　Set the Postprocessing Name to FANUC0i

(4) Press and hold the [Ctrl] key and click all programs under the [BOTTOM] program group. Click Post Process [Postprocess] icon　to open the [Postprocess] dialog box, as shown in Figure 1-139.

(5) Select the postprocessing file [FANUC0i] in [Postprocessor] options, as shown in Figure 1-140, circle 1. Click [Browse for an Output File] icon　under the [Output File] option as shown in Figure 1-140, circle 2, and the [Specify NC Output] dialog box appears. (First create the [nc] folder in disc D) select the D:\nc directory, and enter the file name as "1", as shown in Figure 1-140, circles 3 and 4. Click [OK] to return to the [Postprocess] dialog box, as shown in Figure 1-140, circle 5. Identify the file name location as D:\nc\1 with the file extension .nc, as shown in Figure 1-140, circle 6. Click [OK] to exit the settings, as shown in Figure 1-140, circle 7.

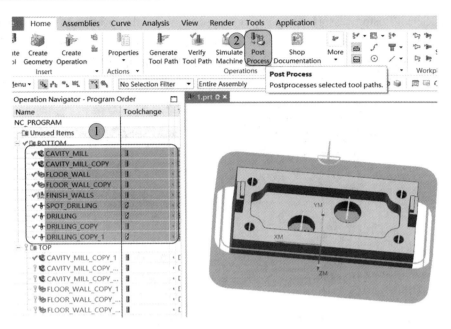

Figure 1-139 Select All Programs under the [BOTTOM] Program Group and Click the [Post Process] Icon

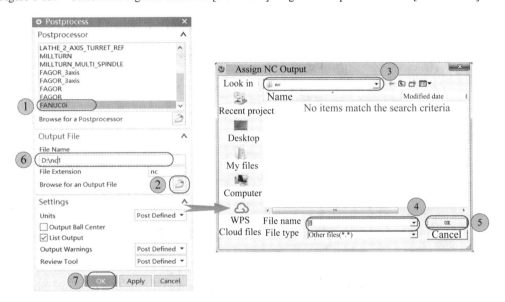

Figure 1-140 Set the Location and Name of Postprocessing File

(6) After clicking [OK] in the step(5), the [Multiple Selection Warning] dialog box pops up, and click [OK] to output all programs to a program group for display, as shown in Figure 1-141.

(7) The G code file pops up and the 1.nc file is generated under the nc folder in disc

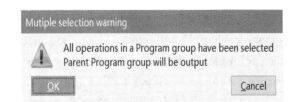

Figure 1-141 Pop-up of [Multiple Selection Warning] Dialog Box

D. If green checkmarks are displayed in front of all programs under the [Bottom] program group, it indicates that the G code file has been generated (note: The machining program that does not generate the G code file is displayed as a yellow exclamation mark), as shown in Figure 1-142.

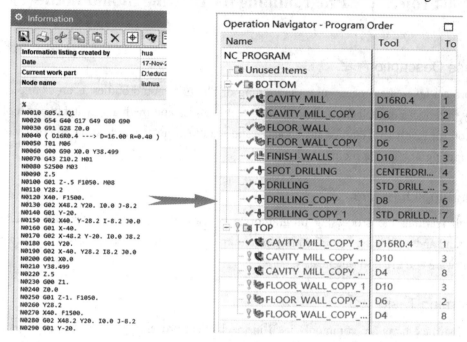

Figure 1-142　Generate a G Code File

(8) Generate machining programs in the above program group in the same way.

Exercises

Complete the programming for after-class exercise files according to the knowledge learned before. Check the 1.prt file in the exercise folder of CD for after-class exercises. Figure 1-143 is the diagram for the exercise.

Figure 1-143　Diagram for Project 1 After-class Exercise

Project V   Programming for Blowing Mould Bottles

➢ Case Description ✍

This project takes the processing programming of blowing mould bottles as an example, which mainly introduces the selection of processing technologies and methods, switching of tool paths, selection of tools of large-scale mould, and cautions in curved surface processing and programming of blowing mould bottles.

➢ Learning Objectives ✍

Through learning the processing programming of blowing mould bottle models, readers can understand and master the manufacturing programming method of triaxial curved surface parts of NX software and use it in other cases.

➢ Learning Tasks ✍

Curved surface parts are commonly used in mechanical processing. Their shapes in processing are complicated and most of them are processed separately. Such parts are widely used in automobile, medical, toy, and mould industry fields.

Taking the processing programming of blowing mould bottles as an example, this case mainly introduces programming of triaxial curved surfaces using NX software. In this process, the key point is to learn and understand the method and application of curved surface processing produces using NX software, and generate manufacturing programs according to different methods and techniques.

5.1 Technological procedures for processing blowing mould bottles

Technological procedures for manufacturing mainly describe each step of the processing process, generally including information about the processed area, processing types (plane milling, curved surface milling, hole machining etc.), description of the procedures, parts clamping, required tools and other necessary information to complete processing.

The overall dimensions of blowing mould bottle are 530 mm × 302.8 mm × 120mm, rectangle, and the material is 45# steel. Before processing mould parts, the forging materials shall be processed through common milling and plain grinding to ensure the overall dimensions meet the drawing requirements, and the six surfaces are in a 90° right-angle relationship. Mould workpiece is mostly processed through single-piece machining instead of batch processing. Besides, milling of steel is time consuming, the cutting resistance force is large, and long-time painting may also influnce the accuracy of the numerically-controlled machine. Therefore, before numerical control machining of a mould part blank, its appearance needs to be processed by using a general-purpose machine tool, and NC machine tool is only used for processing the mould cavity.

5.1.1　Technical analysis of the case

Figure 5-1 shows the part drawing of blowing mould bottles, which requires the processing front cavity part and the left and right positioning edges. Because the overall dimensions of the workpiece are larger than the jaw vise travel range, it can't be fixed by a jaw vice. Therefore, it can only be fixed by a press plate. During this process, the position of the press plate needs to be changed. So during the processing of the workpiece, it needs to be fixed twice.

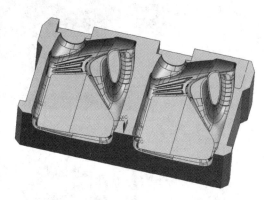

Figure 5-1　Blowing Mould Bottle Model

In order to ensure the dimensional accuracy and tolerance of the processed parts, the processing procedure scheme of the workpiece must be determined before programming.

Process Ⅰ:

Use four press plates to press the middle of the blank and leave the left and right sides for processing. Leave a safe clearance distance between the press plate and tool holder during processing to avoid collisions. The installation position is shown in figure 5-2. Note: Because the workpiece is large and there is much roughing margin, it needs to be fixed with four press plates; otherwise, the workpiece may be moved during milling. After fixation, the first process is to use the milling cutter to process the left and right sides of the workpiece.

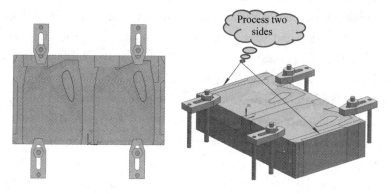

Figure 5-2　Diagram of the Manufacturing Process I

Process II:

After the first processing process, don't rush to move the press plates. Prepare four new press plates and press the four raised parts of the workpiece according to the method shown in figure 5-3. Then remove the four press plates used for fixation in process I. In this way, when the press plates are replaced, the workpiece is still fixed by the four press plates used in process I. After the new press plates are mounted, the workpiece can be fix by them. While the removal of the press plates used in process I will not cause displacement.

Figure 5-3   Diagram of the Manufacturing Process II

5.1.2   Selection of tools for this case

This case is about the processing of blowing mould bottle template. The minimum concave fillet shown in figure 5-5 is semi-finished to 1mm. Due to the high cavity depth, the R1 ball tool can't process to the bottom surface of the mould cavity, we can only use an EDM machine tool to finish the final angle clearance. So we are only going to process it using R3 ball cutter here. During processing of the left and right surfaces of the model, the processing depth of the side surface is large and the side surface is a right-angle surface. If a straight-shank cutter is used, when the cutter moves to a deeper position, the shank may be clamped with scrap iron during milling and damaged. Therefore, in order to avoid clamping, we shall select the milling cutter of which the cutting tip's diameter is ϕ35mm and the shank's diameter is ϕ32mm for processing. During processing of the mould cavity, because the inner contour of the mould cavity is composed of curved surfaces, it shall be machined roughly firstly. Since the workpiece is large and the roughing stock is much, in order to remove the finishing allowance faster during rough machining, we need to select the inserted surface milling cutter D50R0.8 which has a larger diameter. Secondly D12R0.8 inserted drilling and milling cutter is used for secondary roughing to clean the fillet stock left after rough machining. Thirdly, R5 ball cutter is used to roughly process the handle and the upper part of the bottle for the first time to clean the residual part left after processing the last used cutter. Fourthly, we need to use R5 ball cutter to process the curved surface profile of the mould cavity through finish machining. Finally, R3 ball cutter is used to process the fine fillet part in the manufacturing drawing through flow cut and the workpiece manufacturing is completed.

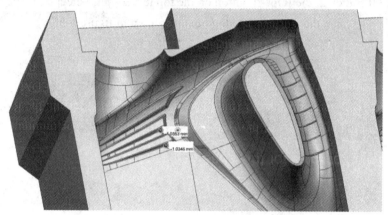

Figure 5-4   Diagram of the Minimum Concave Fillet

The manufacturing technology and process of parts will be arranged rationally according to the processing plan and selection method of tools as mentioned above. The processing technology for this case will be determined following the principle of rough processing first and fine processing later, surface processing first and hole processing later with benchmark unification, as shown in Table 5-1.

Table 5-1   Technical Process Form of Blowing Mould Bottle

Process Number	Sequence Number	Machining Tool	Process Content	Process Name	Tool Name
1	1	Machining center	Rough machining	Deep profile milling	D35R0.8
1	2	Machining center	Finish machining of side wall	Finish milling of wall	D25
2	3	Machining center	Rough machining	Cavity Mill	D50R0.8
2	4	Machining center	Secondary roughing	Milling of the residual part	D12R0.8
2	5	Machining center	Secondary roughing	Milling of the residual part	R5
2	6	Machining center	Finish machining of curved surface	Partial profile milling	R5
2	7	Machining center	Flow cut milling	Flow cut milling	R3
3	8	Machining center	Right mould cavity processing	Parallel translation of all procedures of 2 groups	As above

5.2 Open the model file to enter the processing module

Open the CD attached with the book, open the model file in the case file folder and enter the processing module.

(1) Start NX 12.0 and click the [Open] button on the top left corner. Select 5.prt file from the case file folder in the CD from the "look in" dialog box, as shown in figure 5-5.

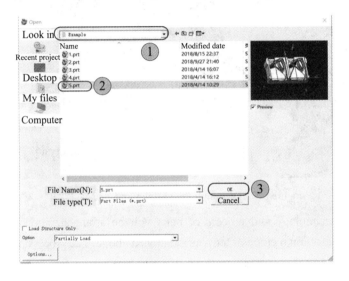

Figure 5-5　Open 5.prt file in the "case" file folder in the CD

(2) Click [Application] button, and then click [Manufacturing] button (or press the shortcut key Ctrl+Alt+M directly) to enter NX12.0 processing module, as shown in figure 5-6.

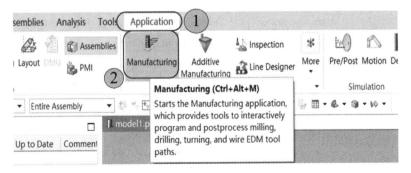

Figure 5-6　Enter NX 12.0 manufacturing module

(3) Select [CAM Session Configuration] [cam_general] by default, and select [mill_contour] in the [CAM Setup to Create] dialog box created. Click [OK] to enter the [Plane Milling] interface, as shown in figure 5-7.

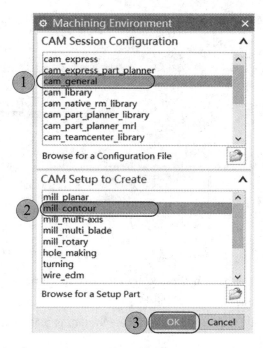

Figure 5-7　Dialog Box for Manufacturing

5.3　Establish parent node groups

The parent node group includes geometry view, machine tool view, program order view and processing method view.

(1) Geometry view: the "manufacturing coordinate system" direction and safety plane can be defined, and geometric parameters such as "part", "blank", and "check" can be set.

(2) Machine tool view: the cutting tools can be defined, you can specify the milling cutter, drill bit, turning tool etc., and save the tool-related data as the default value of the corresponding post-processing command.

(3) Program order view: arrange completed programs in the folder by groups, and arrange machining programs in order from top to bottom.

(4) Processing method view: used to define the types of cutting methods (rough machining, finish machining and semi-finishing). For example, parameters such as "internal tolerance", "external tolerance" and "part stock" are set here.

5.3.1　Create the machine coordinate system

The procedures for creating the machining coordinate system in the [Geometry] view menu are as follows:

(1) Switch [Operation Navigator] to [Geometry], as shown in figure 5-8.

(2) Double-click [MCS_MILL] icon and the dialog box [MCS Mill] pops up, as shown in figure 5-9.

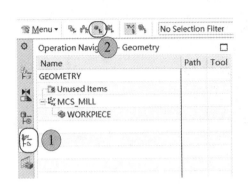

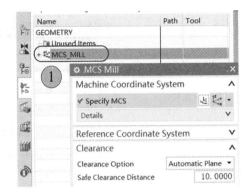

Figure 5-8　Switching of [Geometry] view　　　　Figure 5-9　[MCS Mill] dialog box

(3) As the blank material has been processed by ordinary machine tools in the early stage, the surface shape and overall dimensions of the blank have met the drawing requirements. For the convenience of die assembly, the machine coordinate system is placed in the middle of the blank.

Click ![icon] [Coordinate System] icon, as shown in figure 5-10, circle 1.Select [CSYS of Object], as shown in figure 5-11, circle 2. The function of [CSYS of Object] option is to automatically set the coordinate as the center point of the selected plane. Click the coordinate position in the center point of the upper surface captured automatically from the blank workpiece, as shown in figure 5-12, circle 3. The workpiece is fixed by press plates. When the tool moves left and right, if the safety height of tool lift is too low, the tool may get crashed. Therefore, the safety plane is set to be [Automatic Plane] in the Clearance option, and the safe clearance distance is set to "150mm", which exceeds the press plate height. Finally, click [OK] to exit [MCS Mill], as shown in figure 5-10, circle 4 and circle 5.

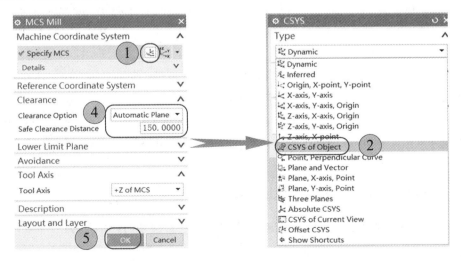

Figure 5-10　Select the [Coordinate System] icon　　　　Figure 5-11　Select [CSYS of Object]

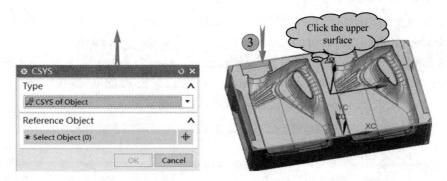

Figure 5-12　Click the box on the upper surface of the workpiece and set the coordinate system

5.3.2 Create the geometry

The procedures for creating the geometry, part blank and geometry check in the [Geometry] view are as follows:

(1) Double-click ⚙WORKPIECE and open [Workpiece] dialog box, as shown in figure 5-13.

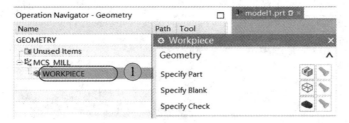

Figure 5-13　Select [WORKPIECE] icon to pop up [Workpiece] dialog

(2) Click 📦 [select or edit the part geometry] icon, as shown in figure 5-14, circle 1. [Part Geometry] dialog box pops up. Click the processed part to make it orange, as shown in figure 5-14, circle 2 and circle 3. Then click [OK] to exit [Part Geometry], as shown in figure 5-14, circle 4.

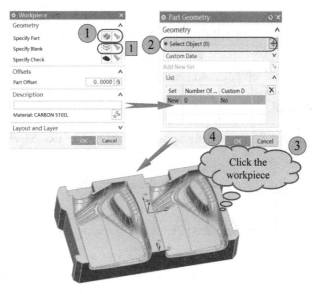

Figure 5-14　Create the [Part Geometry]

(3) Click the ⬡ [Select or edit the blank geometry] icon, as shown in figure 5-14, box 1. The [Blank Geometry] dialog box pops up. Select [Geometry] in the [type] option, as shown in figure 5-15, box 2. Click the blank specified in the figure, as shown in figure 5-15, box 3. Click [OK], as shown in figure 5-15, box 4. Return to [Workpiece] dialog box and then click [OK], as shown in figure 5-15, box 5. Exit [Workpiece] dialog box and finish setting.

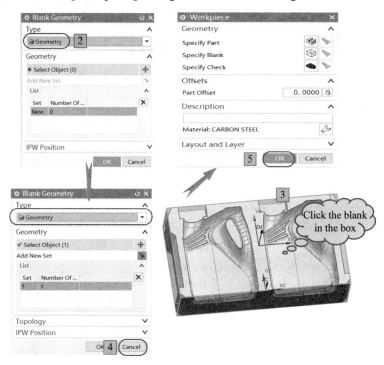

Figure 5-15   Create the [Blank Geometry]

(4) After selecting the blank, the blank box is no longer useful in the editing process. It can be hided to avoid the problem of incorrect operations in the following procedures. The specific method is to click the blank box with the left button and press the shortcut key Ctrl + B on keyboard. If we need to display it back, we can press the shortcut key Ctrl + Shift + K, and select the icon to be restored with a left click, and exit [Display].

5.3.3 Create the tool

The procedures for creating D35R0.8 drilling and milling tool in the [Machine Tool] view are as follows:

(1) Click 🔧 [Machine Tool] icon, and switch the [Operation Navigator] to [Machine Tool], as shown in figure 5-16.

(2) Click the [Create Tool] icon, and the [Create Tool] dialog box pops up, as shown in figure 5-17, circle 2.

(3) Select 🔩 [MILL] icon to create the end mill. Input "D35R0.8" as the name (representing the inserted drilling and milling cutter with a diameter of 35mm and a fillet radius of 0.8mm), as

shown in figure 5-17, circle 3.

Figure 5-16　[Machine Tool] view

Figure 5-17　[Create Tool] dialog box

(4) Set the [Diameter] to "35" and the [Lower Radius] to "0.8". The value of [Tool Number], [Adjust Register] and [Cutcom Register] are all set to "1" (this number represents the tool number, tool radius compensation number, and tool length compensation number. In order to avoid machine collision, it's better to set them the same number). Click [OK] to finish tool creation, as shown in figure 5-18. Click [Shank] to set the [Shank Diameter] to "32mm," [Shank Length] to "150mm", and [Shank Taper Length] to "0", as shown in figure 5-19.

Other tools can be set according to the given parameters before editing the manufacturing program.

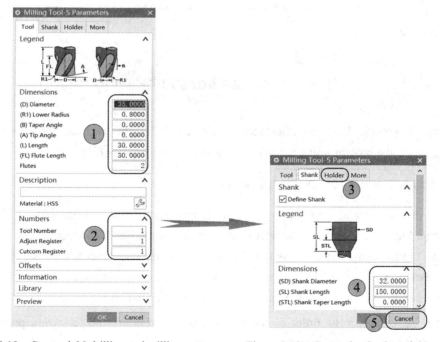

Figure 5-18　Create Φ35 drilling and milling cutter　　　Figure 5-19　Create the shank as Φ32mm

5.3.4 Create the program group

The procedures for creating manufacturing program group file folder in the [Program Order] view are as follows:

(1) Switch [Operation Navigator] to [Program Order], as shown in figure 5-20.

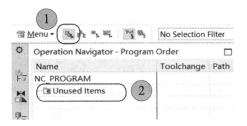

Figure 5-20   Switch [Operation Navigator] to [Program Order]

(2) Double-click [PROGRAM] file folder, as shown in figure 5-20, circle 2. Change the program group name to [1] (or right-click [PROGRAM, and select [Rename] to change the name), as shown in figure 5-21, circle 3.

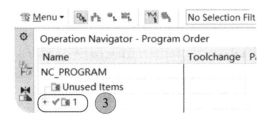

Figure 5-21   Double-click to revise the program group name

(3) Save the file.

5.4   Manufacturing process I of the blowing mould bottle

Manufacturing process I of the workpiece is the left and right surfaces of the model. The following part is the programming method for manufacturing process I.

5.4.1 Programming of rough machining

The width of the left and right surfaces of the maximum margin is 15mm, the width of the machined surface is less than the diameter of the roughing tool. For such workpiece with an open profile and the finishing allowance is less than the tool diameter, the best processing method is ZLEVEL Profile Milling. ZLEVEL Profile Milling can reciprocate manufacturing of open profile and reduce unnecessary tool lifting path, thereby improve the machining efficiency.

In order to better reflect the order of the manufacturing program, the following programming work will be finished in [Program Order].

(1) Click [Create Operation], and the [Create Operation] dialog box pops up, as shown in figure 5-22.

(2) Select [mill_contour] in the pull-down menu of [Type], as shown in figure 5-23. Select 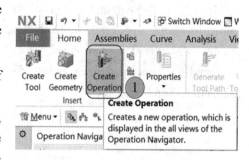 [ZLEVEL Profile Milling], as shown in Figure 5-24, circle 2. Select program group [1] from the pull-down menu, and select [D35R0.8] from the pull-down menu. Select [WORKPIECE] from the pull-down menu of [Geometry], and input a program name according to the manufacturing requirement. In this section, the name will not be amended and the default name will be filled in. Then click [OK], as shown in figure 5-24, circle 3 and circle 4.

Figure 5-22　Click [Create Operation] icon

Figure 5-23　Select [mill_contour]

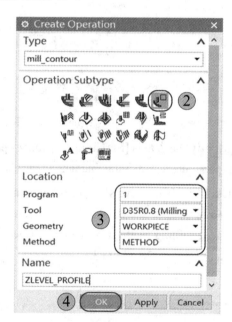

Figure 5-24　Create [Cavity Mill] process

(3) Click [Specify Cut Area] icon in the [ZLEVEL Profile Milling] dialog box popped up, and select the solid face needing manufacturing on the left and right surfaces of the model, as shown in figure 5-25.

(4) Open the drop-down arrow of [Tool] menu, and it shows the tool selected is [D35R0.8] inserted drilling and milling cutter. (Note: this step can be neglected if the preorder is correctly selected.)

Figure 5-25　Dialog box for setting the geometry

(5) Open the pull-down menu of [Tool Axis], and the tool axis shown by default is [+Z Axis], as shown in figure 5-26, circle 4. In the triaxial machining center, the common tool axis used is usually +ZM axis. This item is revised only when a gang tool is used for manufacturing, (Note: This option adopts the default settings for triaxial machining programming.) as shown in figure 5-26, circle 3 and circle 4.

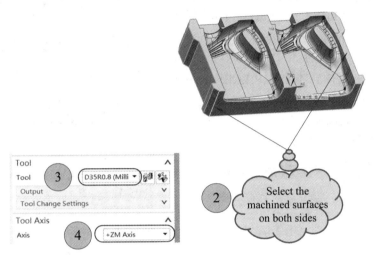

Figure 5-26   [Tool] and [Tool Axis] options

(6) Modify the Z axis [Maximum Distance] parameter to "0.5 mm" each level in [Path Settings], other parameters remaining unchanged, as shown in figure 5-27, circle 1.

(7) Click 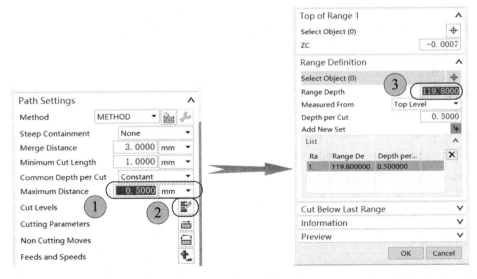[Cut Levels] icon, as shown in figure 5-27, circle 2. Set the [Range Depth] to "119.8mm", make the tool machining to the last tool and table left with a margin of 0.2mm. This can avoid milling of the workbench surface. Click [OK] to exit the setting of [Cut Levels], as shown in figure 5-28, circle 3.

Figure 5-27   0.5mm each level in Z direction   Figure 5-28   Set the scope of machining depth to "119.8"

(8) Click [Cutting Parameters] icon to open the [Cutting Parameters] dialog box, as shown in figure 5-29, circle 1. Set the [Cut Direction] to [Mixed] and [Cut Order] to [Depth First Always] in the [Strategy] label, as shown in figure 5-29, circle 2. These two options can make the cutting path repeated and can reduce unnecessary cutter lifting. Then open the [Stock] menu and set the [Part Side Stock] to "0.2mm", and set the [Intol] and [Outtol] to "0.01mm", as shown in figure 5-29, circle 3 and circle 4. Click [OK] to exit the setting of [Cutting Parameters], as shown in figure 5-29, circle 5.

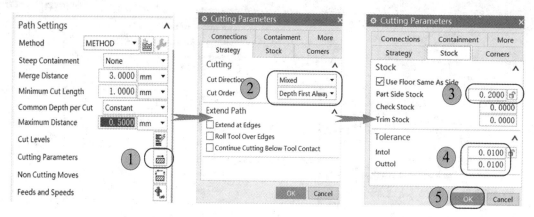

Figure 5-29　Setting of [Cutting Parameters] dialog box

(9) Click [Non Cutting Moves] icon as shown in figure 5-30 to open the [Non Cutting Moves] dialog box.

(10) Set the feeding parameters. First, set the engage parameters in the [Closed Area]. In the mould cavity, the tool usually enters in a helical way. According to the helical engage principle, if the mould cavity size can meet the requirements for helical engage, helical engage is preferred. If the mould cavity size can't meet the requirements for helical engage, select oblique engage or linear engage. The helical engage [Diameter] is "50%" of the tool diameter, the [Ramp Angle] is "5°", the [Height] is "1mm," [Minimum Ramp Length] is "0" (note: if linear engage is available, the minimum ramp length can be "0"; if linear engage is not available, the number shall not be smaller than 50; otherwise, there will be collisions), as shown in figure 5-31, circle 2. In the [Open Area], the engage [Length] is "50%" of the tool, the lifting [Height] is "1mm" to reduce the lifting distance. The specific parameters are set as in figure 5-31, circle 3.

(11) Click the [Transfer/Rapid] label in [Non Cutting Moves] dialog box to set the rapid lifting height, as shown in figure 5-32, circle 1. In order to improve the process velocity and reduce the lifting height, change the lifting height [Within Regions] to [Previous Plane], and set the [Safe Clearance Distance] to "1". The workpiece is fixed with press plates in the middle during manufacturing process I, so the setting method of Previous Plane can't be used. The change tool type [Between Regions] must be [Clearance-Tool], as shown in figure 5-32, circle 2. Otherwise, there may be collisions. Finally, click [OK] to exit the [Non Cutting Moves] dialog box, as shown in figure 5-32, circle 3.

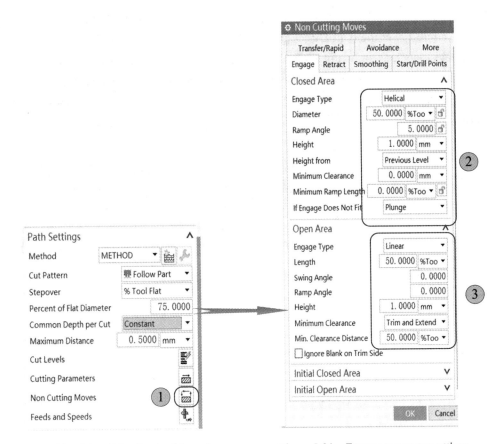

Figure 5-30　Select [Non Cutting Moves]　　　Figure 5-31　Engage parameter settings

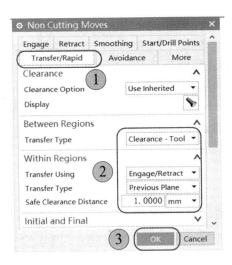

Figure 5-32　Setting of [Transfer/Rapid] Label

(12) Click [Feeds and Speeds] icon and open the [Feeds and Speeds] dialog box and set the [Spindle Speed] to "2000 rpm" (Note: after inputing "2000", remember to click the calculator

icon; otherwise, it will alarm). Enter the cut under the feed rates to be "1500 mmpm", and enter 70% for approach engage. Then click [OK] to exit [Feeds and Speeds] dialog box, as shown in figure 5-33.

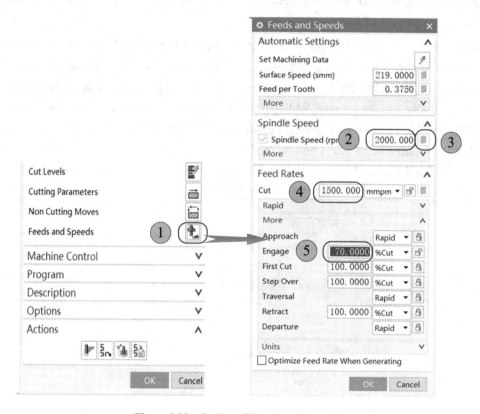

Figure 5-33　Setting of [Feeds and Speeds]

(13) Click [Generate] icon to calculate the manufacturing path. Click [OK] to exit the [ZLEVEL Profile Milling], as shown in figure 5-34.

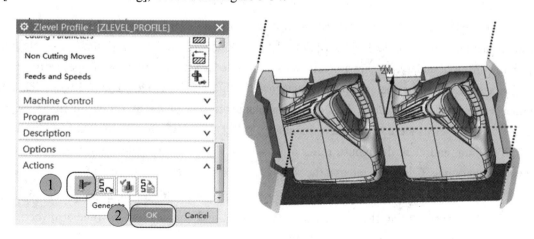

Figure 5-34　Tool Path Generation

5.4.2 Programming of side face finish machining

The programming method with $\phi 25$ alloy tool for finish machining the left and right outer contours of the part is as follows:

(1) Create an alloy tool with a diameter 25 mm, length 150 mm, tool name as D25 alloy, tool No.2. Use $\phi 25$ alloy tool for finish machining of the outer profile. Click [Create Operation] icon and select [mill_planar] in the pull-down menu of [Type] for planar milling, as shown in figure 5-35, circle 1. Select ![icon] [Finish Walls] from [Operation Subtype], as shown in figure 5-35, circle 2. Select [1] from [Program] option, select [D25 Alloy] from [Tool], and select [WORKPIECE] from [Geometry] option, as shown in figure 5-35, circle 3. Finally, click [OK] to enter the [Finish Walls] dialog box, as shown in figure 5-35, circle 4.

Figure 5-35　Create the [Finish Walls] Program

(2) Click ![icon][Specify Part Boundaries] icon and open the [Blank Boundaries] dialog box, as shown in figure 5-36, circle 1. Select [Curve] from Boundaries-Selection Method to create the manufacturing boundary, as shown in figure 5-36, circle 2. Set the [Boundary Type] as [Open], tool side as [Left] and plane as [Specify], as shown in figure 5-36, circle 3 and 4. Right-click the upper surface of the workpiece as the starting manufacturing plane, as shown in figure 5-36, circle 5. Then select the contour curve manually; select the curve from the starting position of climb milling, and click the mouse roller after selecting one boundary line and then select the next boundary line. Click [OK] to exit the setting, as shown in figure 5-36, circle 6, 7 and 8.

(3) Click ![icon][Specify Floor] icon as shown in box 1 of figure 5-36 to pop up the [Plane] dialog box. Set the manufacturing floor, select the model floor, and double-click the blue arrowhead to make it upward, and enter "0.2mm". Lift the final cut for 0.2mm to avoid milling the workbench, as shown in Figure 5-37.

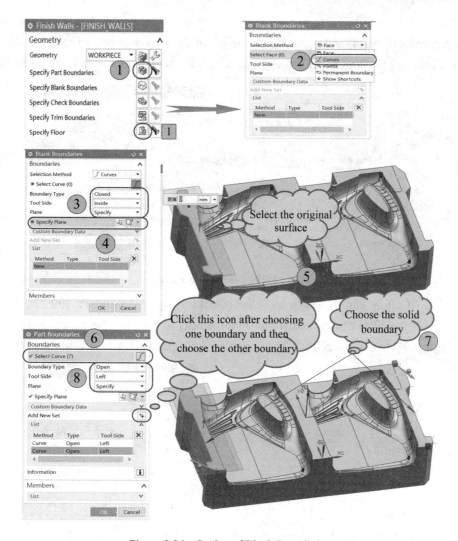

Figure 5-36   Setting of Blank Boundaries

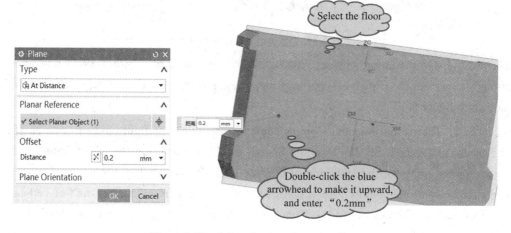

Figure 5-37   Select the manufacturing floor

(4) Click ▤ [Cut Levels] icon to pop up the [Cut Levels] dialog box. Select [Constant] from the pull-down menu of [Type], and make the cut height of each layer to be constant. Set the [Depth Per Cut] [Common] to "20mm", and click [OK] to exit the setting, as shown in figure 5-38.

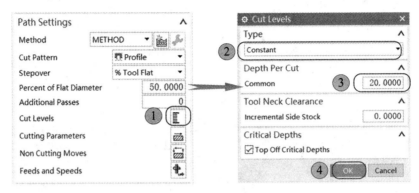

Figure 5-38　Set the [Cut Levels]

(5) Click ⛏ [Cutting Parameters] icon, select [Stock] from the [Cutting Parameters] dialog box popped up. Set the [Part Stock] and [Final Floor Stock] to "0", set the [Intol] and [Outtol] to "0.01". Click [OK] to exit the setting. The operation procedures are provided in figure 5-39.

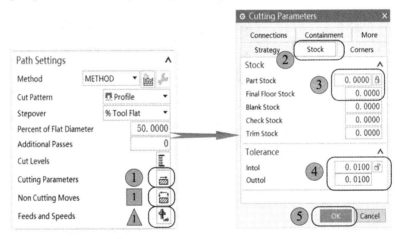

Figure 5-39　Set the [Stock] and [Tolerance]

(6) Click ⛏ [Non Cutting Moves] icon, as shown in box 1 of figure 5-39 to pop up [Non Cutting Maves]. Set the [Engage Type] to [Linear] and turn on the radius compensation function, as shown in figure 5-40.

(7) Click ⛏ [Feeds and Speeds] icon, as shown in triangle 1 of figure 5-39 to pop up [Feeds and Speeds] dialog box. Set the [Spindle Speed] to "2000 rpm" and the [Feed Rates] [Cut] to "500 mmpm" (during finish machining of the side wall, the tool shall not move too fast, and the speed shall be controlled below "500 mmpm". Otherwise, large roughness of the finished surface will affect the manufacturing quality). Then click [OK] to exit the [Feeds and Speeds] dialog box, as shown in figure 5-41.

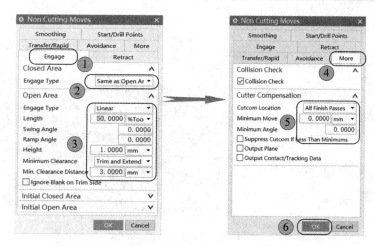

Figure 5-40　Setting of [Non Cutting Moves] dialog box

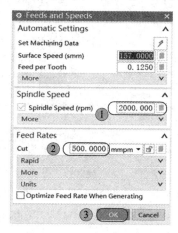

Figure 5-41　Setting of spindle speed and feed speed

(8) Click [Generate] icon to calculate the manufacturing program, and generate [Finish Walls] for finish machining of the side wall paths. Click [OK] to exit the setting of [Finish Walls], as shown in figure 5-42.

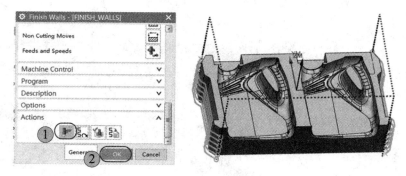

Figure 5-42　Manufacturing program to generate [Finish Walls]

(9) Simulate the manufacturing program edited. Click [Program Order] icon on the upper left, and select all manufacturing programs having been edited on the interface of [Program Order], as

shown in Figure 5-43.

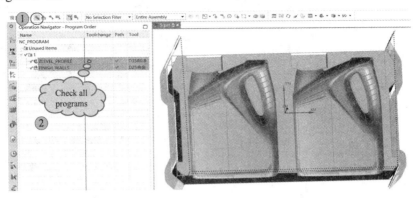

Figure 5-43   Check all manufacturing programs on the [Program Order]view

(10) Click [Home] [Verify Tool Path] icon and open the [Tool Path Visualization] dialog box, as shown in figure 5-44.

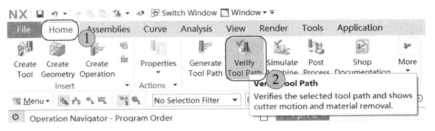

Figure 5-44   Click [Verify Tool Path] icon

(11) Click [3D Dynamic] to switch the [Simulation Animation] to [3D Stereoscopic Model]. Then click ▶ [Play] icon at the bottom to finish the manufacturing simulation of the floor path of the workpiece, as shown in figure 5-45.

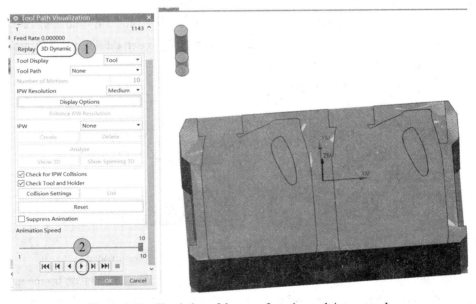

Figure 5-45   Simulation of the manufacturing path in process I

5.5 Manufacturing process II of the blowing mould bottle

After manufacturing process I, fix the press plates on the left and right sides and leave the middle part of the mould cavity according to the above technical analysis for process II milling. Before editing the manufacturing process II, create a new program group [2] in the [Program Order] view.

5.5.1 Programming of rough machining

First, use the Cavity Mill processing method to finish the rough machining programming of the part. In order to better reflect the sequential order of the manufacturing program, we use the [Program Order] in the following programming.

(1) Create a 50mm diameter, 0.8surface milling tool with a diameter of $\phi50$ and a fillet radius of R0.8, and a length of 100mm for rough machining. Enter the name "D50R0.8", and tool number 3#.

(2) Create [Cavity Mill] program, select [mill_contour] curved surface milling from the pull-down menu of [Type], as shown in figure 5-46. Click [Cavity Mill] icon, and the newly-created [1] program group from the pull-down menu of program, select surface milling tool [D50R0.8] from the pull-down menu of [Tool], and select [WORKPIECE] from the pull-down menu of [Geometry], and then click [OK], as shown in Figure 5-47.

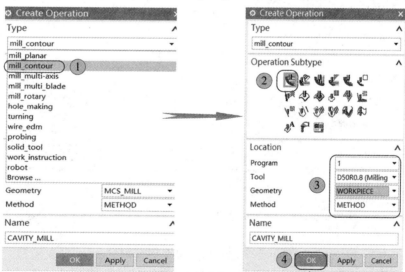

Figure 5-46　Select [mill_contour]　　　　Figure 5-47　Create [Cavity Mill] Process

(3) Select [Specify Cut Area] icon from the [Cavity Mill] dialog box popped up, and select the single cavity curved surface. The model is large and it has many curved surfaces, so if two cavity curved surfaces are selected for manufacturing at the same time, the calculation workload will be large and the calculation speed will be slow. In order to improve the calculation speed of the

program, in this case, only one cavity program will be made, and the program will be transferred horizontally for making the program of the other cavity, as shown in figure 5-48.

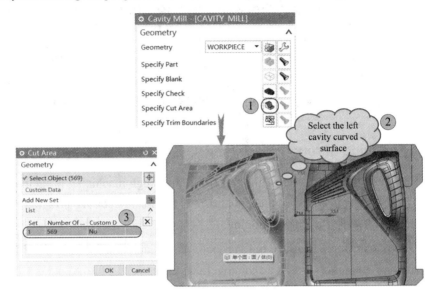

Figure 5-48　Select the curved surface of the left cavity as the cut area

(4) [Path Settings] is the main content of [Cavity Mill] parameter setting. Since the cavity curved surface manufactured is an open contour, [Follow Part] cut mode is more suitable. Therefore, the cut [Pattern] can be set as [Follow Part]. The [Percent of Flat Diameter] of stepover in direction is "75%"; the [Maximum Distance] of level in Z direction is "0.5mm" each layer. The operation procedures are shown in figure 5-49.

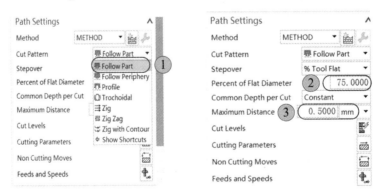

Figure 5-49　[Path Settings] parameters

(5) Click [Cutting Parameters] icon to open the [Cutting Parameters] dialog box. Set the [Cut order] as [Depth First] in the [Strategy] label. Select [Stock] label and set [Part Side Stock] to "0.2mm". Select [Connections] label. Since the [Cut Pattern] is [Follow Part], there will be [Open Passes] in the [Connections] label. Here, we select [Alternate Cut Direction]. After the setting, click [OK] to exit [Cutting Parameters], as shown in figure 5-50.

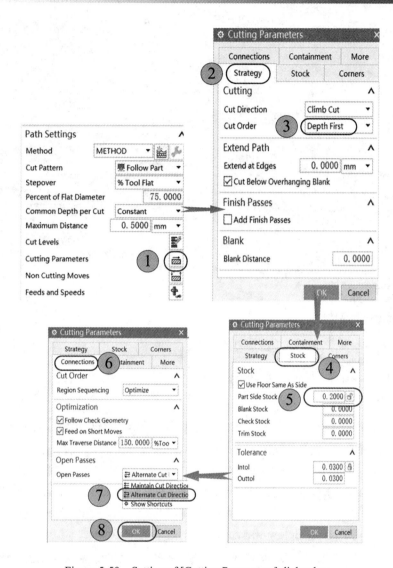

Figure 5-50　Setting of [Cutting Parameters] dialog box

(6) Click [Non Cutting Moves] icon and enter the [Non Cutting Moves] dialog box to set the engage parameters. First, set the engage parameters in the [Closed Area]. In the mould cavity, the [Engage Type] usually enters [Helical]. The bottom center of the surface milling tool has no blade, so it has no cutting ability. Therefore, surface milling tool can't be used in a linear way, and the semi-diameter for helical engage shall be large. Therefore, when setting helical engage, the engage [Diameter] shall be 100%; the [Ramp Angle] is "5°", the [Height] is "1mm", and the [Minimum Ramp Length] is "100" (note: the surface milling tool can't be engaged in a linear way, so the number shall be larger, in case of linear engage and collisions). In the [Open Area], the engage [Length] is "50%" of the tool length, the lifting [Height] is "1mm" to reduce the lifting distance. The specific parameters are set as shown in Figure 5-51.

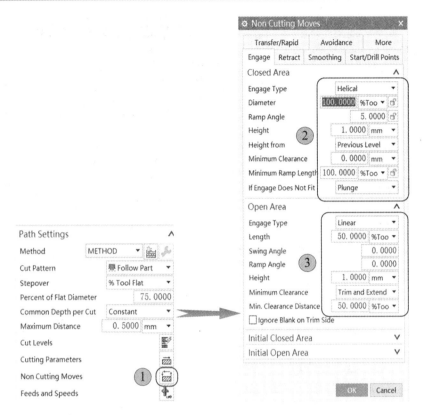

Figure 5-51　Engage parameters for [Non Cutting Moves]

(7) Click the [Transfer/Rapid] label on the top to set the rapid lifting height. In order to improve the process velocity and reduce the lifting height, change the [Tansfer Type] in [Between Regions] and [Within Regions] to [Previous Plane], and set the [Safe Clearance Height] to "1mm". In this way, it can reduce unnecessary cutter lifting and save the process time. However, in most of the time, rapid moving happens below the zero plane of the workpiece, so the G00 motion of the machine shall be moving along the shortest distance between two points, instead of fold line between two points; otherwise, there will be collisions. Before manufacturing, ensure to enter the oblique line moving mode of G00 under MDI and observe the moving mode of the machine. If it's wrong, revise the machine parameters, or set the [Transfer Type] to [Clearance-Tool] according to the initial setup of NX12.0 [Transfer/Rapid]. Return to the safety plane each time in lifting, so as to avoid impacts and increase the process time. Finally, click [OK] to exit the dialog box of [Non Cutting Moves], as shown in Figure 5-52.

(8) Click ![icon] [Feeds and Speeds] icon to open the

Figure 5-52　Setting of [Transfer/Rapid] label

[Feeds and Speeds] dialog box and set the [Spindle Speed] as "1000 rpm" (note: after entering 1000, remember to click the calculator icon; otherwise, it will give out an alarm). Enter the [Cut] under the [Feed Rates] to "800 mmpm", and enter "70%" for [More] [Engage]. Then click [OK] to exit the [Feeds and Speeds] dialog box, as shown in figure 5-53.

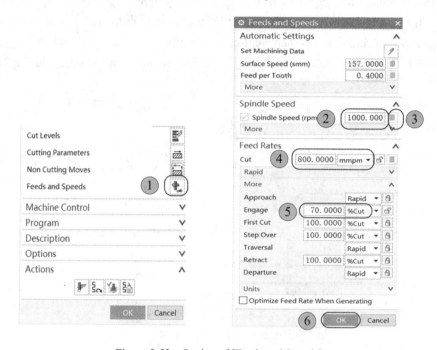

Figure 5-53　Setting of [Feeds and Speeds]

(9) Click ⊩ [Generate] icon to calculate the manufacturing path. Click [OK] to exit the [Cavity Mill], as shown in figure 5-54.

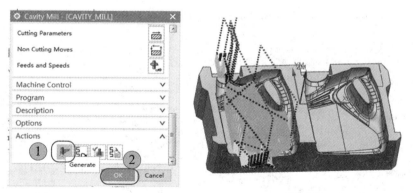

Figure 5-54　Tool path generation

5.5.2　Programming for secondary roughing

Since the tool diameter in the first rough machining is large, many parts can't be manufactured. Therefore, the tool with a smaller diameter shall be used for the secondary rough machining. Here we select the drilling and milling cutter with a diameter of Φ12mm for the secondary roughing of the cavity.

(1) Click the [Machine] view to create a drilling and milling tool with a diameter of 12mm and a smaller radius of 0.8. Select the tool number to be 4#, and the tool name to be D12R0.8.

(2) Use the residual milling method for secondary roughing programming of the cavity. Right-click the [CAVITY_MILL] program that has been made in the [Machine] view, and select [Copy], as shown in figure 5-56. Then right-click [D12R0.8] and select [Paste Inside], as shown in Figure 5-56. Thus a new wrong [Cavity Mill] program is generated, as shown in Figure 5-57.

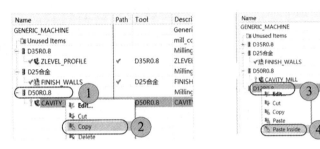

Figure 5-55　Copy [Cavity Mill] program　　Figure 5-56　Paste Inside [Cavity Mill] program

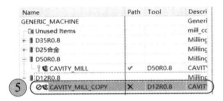

Figure 5-57　New [Cavity Mill] program

(3) Open the newly copied [Cavity Mill] program with a double-click, as shown in Figure 5-57, circle 5. Set the [Maximum Distance] in Z direction to "0.2mm", as shown in Figure 5-58.

(4) Click [Cutting Parameters] icon to pop up the [Cutting Parameters] dialog box, as shown in Figure 5-59.

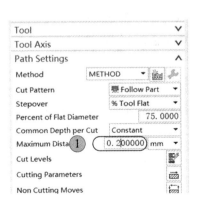
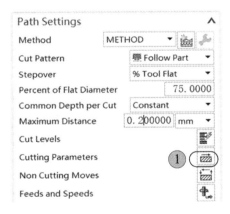

Figure 5-58　Set the [Maximum Distance] to "0.2mm"　　Figure 5-59　Click [Cutting Parameters] Icon

(5) Select [Containment] label in the [Cutting Parameters] dialog box to set the residual blank of

the workpiece. Select [Use Level Based] from the pull-down menu of [In Process Workpiece] and set the blank processed this time to the residual part of the last time. In [Residual Mill], this option is [Use Level Based] by default. The only difference between [Cavity Mill] and [Residual Mill] is this option. Then click [OK], as shown in Figure 5-60.

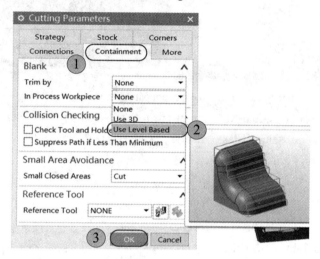

Figure 5-60　Setting of [Containment] Label

(6) Click the ![Feeds and Speeds] icon to set the [Spindle Speed] to "4000 rpm." Click ![calculator] icon on the right side of the speed, and finally click [OK] to exit [Feeds and Speeds] dialog box, as shown in Figure 5-61.

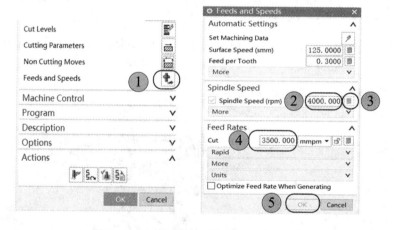

Figure 5-61　Setting of [Feeds and Speeds]

(7) Click ![Generate] icon to calculate the manufacturing program and generate [Cavity Mill] manufacturing path. Then click [OK] to exit the [Cavity Mill] setting, as shown in Figure 5-62.

(8) Use R5 ball cutter for rough machineing of the residual part of D12R0.8. First, click the [Machine] view to create a ball cutter with a diameter of 10mm. Select the tool number as 5#, and the tool name R5.

 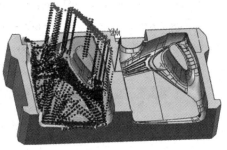

Figure 5-62　Generate the tool path for secondary roughing

(9) Right-click the last [Residual Mill] program in the [Machine] view, select [Copy], and then right-click [R5] tool to select [Paste Inside]. Thus a new wrong [Cavity Mill] program is generated, as shown in Figure 5-57.

(10) Double-click to open the copied [Cavity Mill] program. Select [Constant] in the [Stepover]; set [Maximum Distance] in XY direction to "0.3", set the [Maximum Distance] in Z direction to "0.2mm", as shown in Figure 5-63.

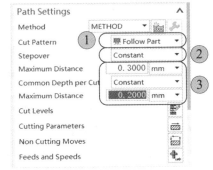

(11) Click [Feeds and Speeds] icon to set the [Spindle Speed] to "8000 rpm". Click the [calculator] icon on the right side of the speed, and finally click [OK] to exit the [Feeds and Speeds] dialog box, as shown in Figure 5-61.

Figure 5-63　Setting of cut levels in XY and Z directions

(12) Click [Generate] icon to calculate the manufacturing program and generate [Cavity Mill] manufacturing path. Click [OK] to exit the [Cavity Mill] setting, as shown in figure 5-64.

 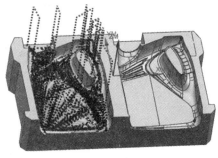

Figure 5-64　Generate the tool path for secondary roughing

5.5.3　Programming of finish machining of curved surface

The commonly used program for finish machining of curved surface is [Fixed Axis Contour Mill]. The new edition of the NX software can set different processing methods according to the different oblique angles of the curved surface, namely [Steep and Non-steep] in NX. Generally, the principle for programming finish machining of curved surface is that isometric processing method can be used for steep surfaces to be manufactured from top down level by level, and ZigZag or circulating method can be used for finish machining of non-steep surfaces. In the following part, we will introduce the programming for finish machining of curved surfaces.

(1) Click [Create Operation] icon to pop up the [Create Operation] dialog box. Select [mill_contour] in the pull-down menu of [Type]. Select [Contour_Area] icon, select [2] program group from the pull-down menu of [Program], select ball cutter [R5] from the pull-down menu of [Tool], and select [WORKPIECE] from the pull-down menu of [Geometry], and then click [OK], as shown in Figure 5-65.

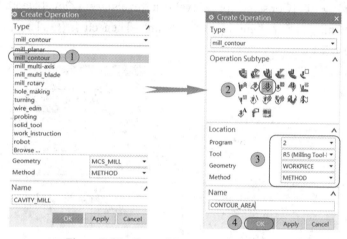

Figure 5-65　Create [Contour_Area] Process

(2) Click [Specify Cut Area] icon in the [Contour_Area] dialog box popped up. Select the left single cavity curved surface and click [OK] to exit [Cut Area] dialog box, as shown in Figure 5-66.

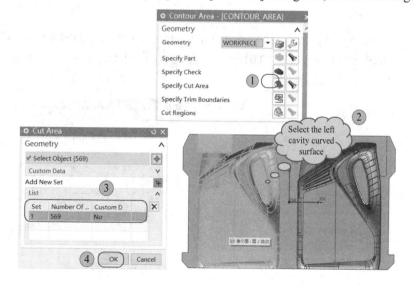

Figure 5-66　Select the curved surface of the left cavity as the cut area

(3) Click 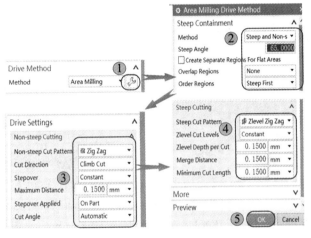[Editing] icon on the right side in [Drive Method] option to open the [Area Milling Drive Method] dialog box. Select [Steep and Non-steep] from the [Method] option of [Steep Containment]. Set the advance mode of tool path as contour manufacturing form for 65°—90° curved surface. Set the ZigZag manufacturing form for 0—65° curved surface manufacturing. Set the [Non-steep Cut Pattern] as [ZigZag], change [Stepover] to [Constant], and the [Maximum Distance] between cuts is "0.15 mm". (Note: During finish machining of the curved surface, if the stepover is too large, the smooth finish of the curved surface is poor; if the stepover is too small, the smooth finish is good but the manufacturing time is long. Years of manufacturing experience shows that the effect is the best when the stepover is 0.15mm). Select [On Part] for "Stepover Applied", and in this way, the curved surface effect is good. Select [Zlevel ZigZag] for [Steep Cut Pattern]; select [Constant] for [Zlevel Cut Levels]; set [Zlevel Depth Per Cut], [Merge Distance], and [Minimum Cut Length] to "0.15mm". Then click [OK] to exit the setting, as shown in Figure 5-67.

Figure 5-67　Setting of [Area Milling Drive Method] dialog box

(4) Click 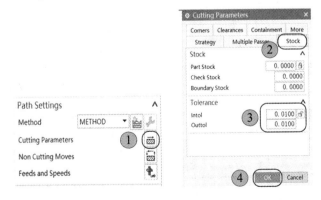[Cutting Parameters] icon to open the [Cutting Parameters] dialog box. Click [Stock] label to set the [Intd] and [Outtol] to "0.01mm", and then click [OK] to exit [Cutting Parameters] dialog box, as shown in Figure 5-68.

Figure 5-68　Set [Intol] and [Outtol] to "0.01mm"

(5) Click  [Feeds and Speeds] icon to open [Feeds and Speeds] dialog box and set the [Spindle Speed] to "10000rpm." Enter the [Cut] under the [Feed Rates] to "1500mmpm", and enter "70%" for [Approach Engage]. Then click [OK] to exit the [Feeds and Speeds] dialog box, as shown in Figure 5-61.

(6) Click [Generate] icon to calculate the manufacturing path. Click [OK] to exit the [Contour Area], as shown in Figure 5-69.

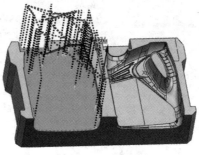

Figure 5-69　Tool Path Generation

5.5.4　Programming for flow cut of finish machining of curved surface

Use R3 ball cutter to edit the curved surface flow cut program. The engage method of curved surface flow cut is similar to that of finish machining. It's also can be divided into Steep and Non-steep manufacturing according to the angle. The following is the setting method of curved surface flow cut program.

(1) Click the [Machine] view, create a ball cutter with a diameter of 6mm. Select the tool number to 6#, and the tool name R3.

(2) Use flow cut milling to finish the finish machining of the small fillet of the curved surface. Right-click the newly made [CONTOUR_ AREA] program in the [Machine] view, and select [Copy]. Then right-click [R3] tool and select [Paste Inside]. Thus a new wrong [Contour_Area] program is generated, as shown in Figure 5-70.

Figure 5-70　Copy the new [Contour_AREA] program

(3) Click to open the pull-down menu of [Drive Method] to select the [Flow Cut] method and enter [Flow Cut Drive Method] dialog box. Select the [Cut Pattern] to [Reference Tool Cut] from the pull-down menu of [Flowcut Type]. Reference Tool Cut is to automatically calculate the residual fillet part for flow cut according to the diameter of the reference tool determined in the program. Select [ZigZag] cut pattern from the pull-down menu of [Non-steep Cutting], and set the [Stepover] to "0.15 mm", and the [Sequencing] is [Steep First], namely manufacturing of the steep surface first and non-steep surface later. Such method can prolong the tool life and avoid collisions

and folding of tool. Select [Crosscut ZigZag] for [Steep Cut Pattern]. The path of Crosscut ZigZag is similar to ZLevel Profile Milling. It's also level-by-level manufacturing from top down. Select [High To Low] for [Steep Cut Direction] and set the [Stepover] to "0.15mm". Finally, open the [Reference Tool] setting and set the [Reference Tool] to the last milling tool [R5 milling tool-ball cutter]. Click [OK] to exit the flow cut setting. The specific setting method is shown in Figure 5-71.

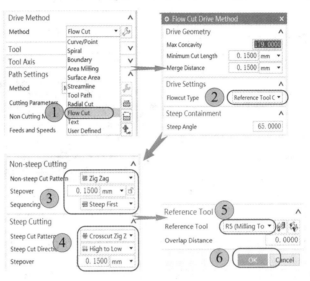

Figure 5-71   Setting of [Flow Cut Drive Method] dialog Box

(4) Click 🏳 [Generate] icon to calculate the manufacturing path. Click [OK] to exit the [Contour Area] dialog box, as shown in Figure 5-72.

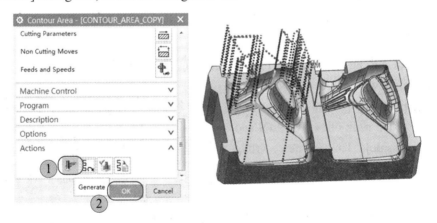

Figure 5-72   Tool Path Generation

5.6   Change the tool path to make the other cavity program

Translate the tool path horizontally to make the manufacturing program of the right cavity. The specific operation method is shown as below.

(1) Click 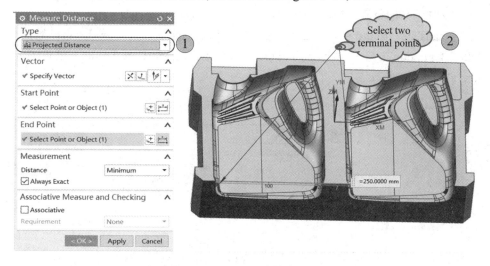[Program Order] icon to switch the [Navigator] to [Program Order] view. Then click the [Create Program] icon on the top left corner of the screen to pop up the [Create Program] dialog box and enter name [3]. Click [OK] to pop up the [Program] dialog box and then click [OK] to exit [Program] dialog box. Add a [3] program group under the program order navigator, as shown in Figure 5-73.

Operation Navigator - Program Order			□
Name	Toolchange	Path	Too
NC_PROGRAM			
Unused Items			
✓ 1			
✓ ZLEVEL_PROFILE		✓	D35
✓ FINISH_WALLS		✓	D25
2			
CAVITY_MILL		✓	D50
CAVITY_MILL_COPY		✓	D12
CAVITY_MILL_COPY...		✓	R5
CONTOUR_AREA		✓	R5
CONTOUR_AREA_C		✓	R3
+ 3			

Figure 5-73   Set a new program group [3]

(2) Click [Measure] icon and select [Projected Distance], and select CSYS X-direction. Select two terminal points at the same position on two cavities respectively, and the center distance measured between two cavities is 250mm, as shown in Figure 5-74, circle 2.

Figure 5-74   The center distance measured between two cavities is 250mm

(3) Check all manufacturing programs under program group [2], as shown in Figure 5-75.

(4) Right-click [Object] and then click [Transform] to pop up the [Transform] dialog box, as shown in Figure 5-76.

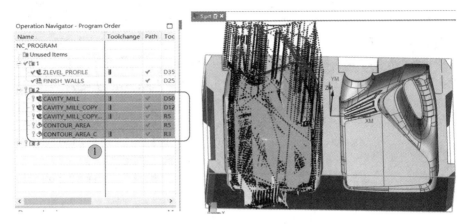

Figure 5-75   Check all programs under program group [2]

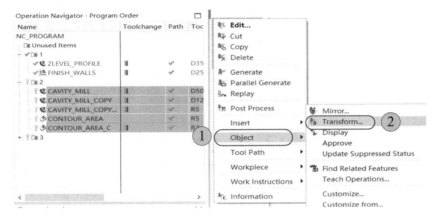

Figure 5-76   Select the [Transform Directive] with the right button

(5) Select [Translate] from the pull-down menu of [Type], and enter center distance "250mm" to the [Delta XC] option in [Transformation Parameters], and select [Copy] radio button. Click [OK] to exit, as shown in Figure 5-77.

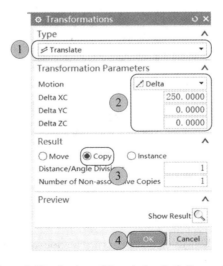

Figure 5-77   Setting of Translation Path Parameter

(6) Drag the newly generated right cavity program to program group [3], as shown in Figure 5-78.

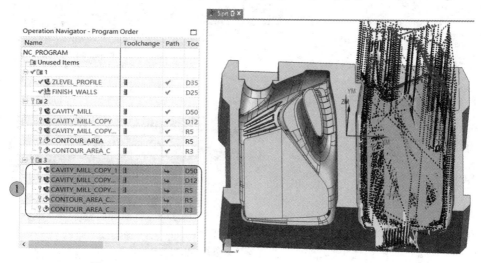

Figure 5-78 Generate the right cavity manufacturing program

(7) Simulate all manufacturing programs finished. Click ⬚ [Program Order] icon on the upper left and select all finished manufacturing programs on the program order interface. Click the [Verify Tool Path] icon on the [Home] page and open the [Tool Path Visualization] dialog box. Click [3D Dynamic] to switch the simulation animation to be 3D stereoscopic model. Then click ▶ [Play] icon at the bottom to finish the manufacturing simulation of the floor path of the workpiece, as shown in Figure 5-79.

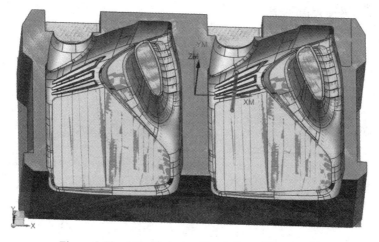

Figure 5-79 Manufacturing Simulation of Process Ⅱ

5.7 Generate G-code file

(1) Press [Ctrl] on the keyboard, and left click to select all programs under program group [1]. Click [Postprocess] icon to pop up the [Postprocess] dialog box, as shown in Figure 5-80.

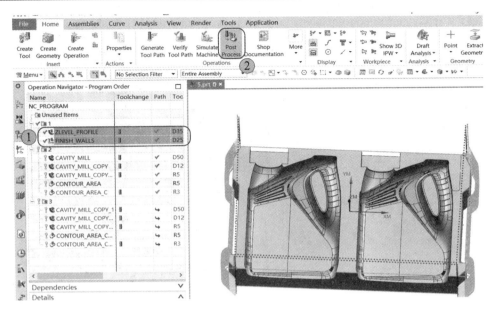

Figure 5-80   Select all programs under program group [1] and click [Postprocess] icon

(2) Select [FANUC0i] postprocess file from [Postprocessor], click ⟨⟩[Browse for an Output File] under [Output File] to pop up [Specify NC Output] dialog box. (First, create [nc] file folder in D disk.) Select D:\nc, enter the file name [1], and click [OK] to return to the [Postprocess] dialog box. Confirm the file name position is D:\nc\1, and the filename extension is ".nc". Click [OK] to exit the setting, as shown in Figure 5-81.

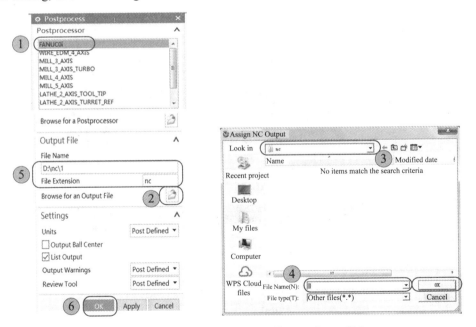

Figure 5-81   Setting of Postprocess File Location and Name

(3) Click [OK] to pop up the [Multiple selection warning] dialog box, and click [OK] to output all programs under one program group, as shown in Figure 5-82.

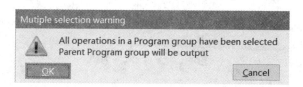

Figure 5-82 [Multiple selection warning] dialog box pops up

(4) G-code file pops up and 1. nc file under nc file folder of D disc is generated. After a G-code file is generated, all programs under the bottom program group will have a green check mark, showing that G-code file has been generated. In front of the manufacturing program generating no G-code file, there will be a yellow exclamation mark, as shown in Figure 5-83.

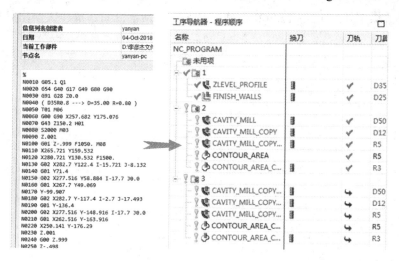

Figure 5-83 Generate G-code file

(5) Generate the process programs of the other two program groups in the same way.

Exercise

Finish the programming in the exercise according to what you have learnt in this class. The exercises can be found in 5.prt file in the "exercise" file folder in the CD Figure 5-84 is the figure for After-class exercise.

Figure 5-84 Figure for Project 5 After-class Exercise

参 考 文 献

[1] 康亚鹏，杨小刚，左立浩. UG NX 8.0 数控加工自动编程[M]. 4 版. 北京：机械工业出版社，2017.

[2] 袁锋. UG CAM 数控自动编程实训教程[M]. 北京：机械工业出版社，2015.

[3] 陈桂山，贾广浩，李明新. UG NX 8.5 数控加工入门与提高[M]. 北京：机械工业出版社，2013.

[4] 贺建群，徐宝林. UG NX7.0 多轴加工实例教程[M]. 北京：清华大学出版社，2011.